ENCYCLO

2

CONTEMPORARY STUDIES

Encyclopedia of the Mediterranean
is published jointly by the following publishers
in their respective languages:

Alif-Editions de la Méditerranée, Tunisia
CIDOB-Icaria, Spain
Dar el-Ferjani, Libya
Edisud, France
Editoriale Jaca Book, Italia
Les Editions Toubkal, Morocco
Midsea Books, Malta

BERNARD KAYSER

THE MEDITERRANEAN - GEOGRAPHY OF THE FRACTURE

Translated from the Italian by
LOUIS J. SCERRI

Midsea Books
Malta
1998

Published for the first time in 1998
Printed in Malta

General Editor: Louis J. Scerri

Production:
Mizzi Design & Graphic Services

The Encyclopedia of the Mediterranean,
is promoted by an internation association SECUM
- Sciences, Education et Culture en Méditerranée
with offices in Aix-en Provence, Tunisia, Casablanca,
Milan, and Malta.

ISBN: 99909-75-61-2

Further information on this and
other publications may be obtained from
Midsea Books Ltd
Tower Building, Sulphur Lane, Blata l-Bajda HMR 02, Malta
Tel. (+356) 237617, Fax: (+356) 237643
e-mail: kkm@maltanet.net

INDEX

THE MEDITERRANEAN - GEOGRAPHY OF THE FRACTURE

The Mediterranean is a fiction and an image. It is not a geography; as Predag Matvejić suggests: 'It is impossible to explain what drives us unceasingly to recompose the Mediterranean mosaic.'

Still the image the Mediterranean offers us is not simply a virtual one. Its reality is confirmed by a number of maps. First of all by that which shows the diffusion of the olive tree and those plants which normally flourish with it; such a map traditionally provides a biogeographic parameter, namely that of the Mediterranean basin. Then there is the map which shows the extent of the Roman Empire all around the *mare nostrum*. This goes back to the early years of the first millennium and identifies the area of a stable civilization which can be described as 'Mediterranean'. Finally, there are the colour postcards; all along the coast and in the hinterlands laid waste by modern man, photographers and tourists can still avail themselves

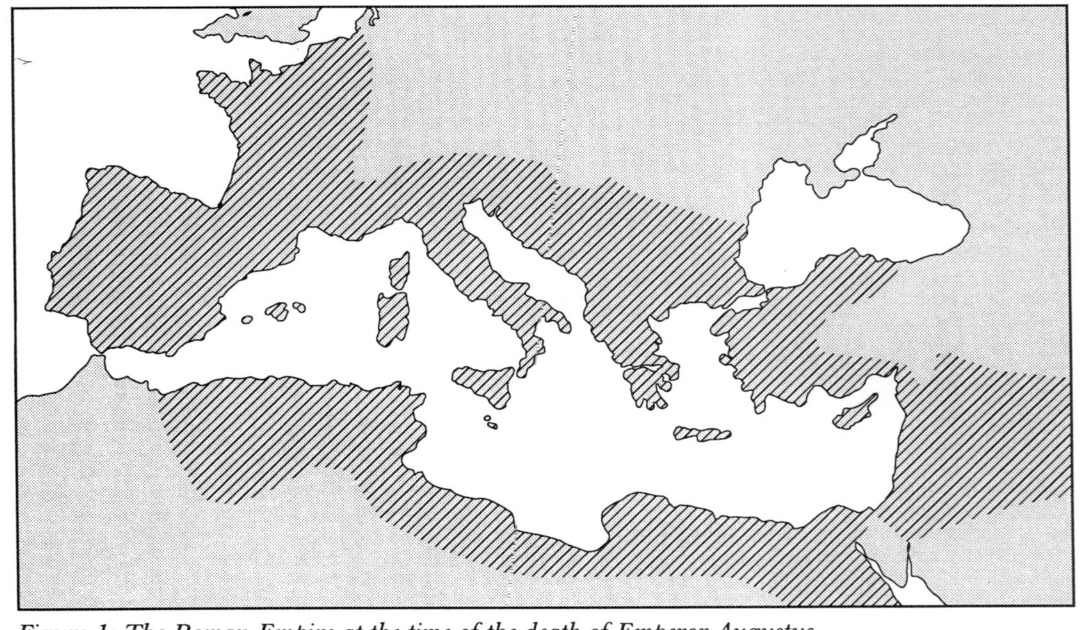

Figure 1: The Roman Empire at the time of the death of Emperor Augustus

of an unequalled natural light and carry back with them mementoes of bucolic landscapes which recall the literary descriptions of yore.

The Mediterranean imposes itself. Nowhere else, for such a large area with so great diversities, can a unity be perceived so intensely. Today, in fact, for the people of the western hemisphere and for their media of communication, the Mediterranean offers quite a definite image; an idyllic place for the sun and the sea and bathed in a Graeco-Latin aura: an immense area destined for holidays and sloth.

Its boundaries are, however, not circumscribed by time or place. 'One cannot understand', continues Matvejić, 'how to determine them, nor with effect to what. They are neither decided by history, nor ethnic groups, nor nations, nor states. It is a chalk circle which is being drawn and rubbed out all the time.' If one uses the term 'Mediterranean' for those countries around the coast, the ambiguity of the definition makes itself felt, particularly in the political sphere. Beyond the coasts, the mountains, and the deserts, do the industrial regions of the north belong to a single entity or to a community? The Mediterranean is a community on paper. Are the Mediterranean countries, therefore, only those countries which have their own interests - so often conflicting ones - in the sea that bathes their coasts?

Anyway, the Mediterranean and the subjects connected with it are inseparable. In 1864, Elisée Reclus, the distinguished geographer who was so misunderstood in his own times, stressed 'the particular fascination which the Mediterranean has for all those who live in the inner reaches of the continent', and foresaw that gradually seasonal masses of populations would be attracted to its shores and would transform 'its cities into enormous caravanserais'. His words were echoed by Jules Sion in 1934 in his *Universal Geography*: 'Whoever goes away from its shores will always retain a nostalgic memory.' André Siegfrid, on his part, in his *Vue Générale* of 1943, attempted to provide an objective basis for these impressions, even going to the length of asserting - he even described the existence of a 'Mediterranean race' - that the Mediterranean was 'everywhere the same; the differences are less important than the similarities ... One always knows, to within a kilometre or so, if one is in a Mediterranean region or not.'

Do we, therefore, have to keep pushing ourselves in a search for absolute terms? Albert Camus, touched with the sea's magic spell, suggests: 'Every time that a particular doctrine encountered the Mediterranean basin, the Mediterranean always managed to remain intact in the clash of concepts that resulted, with the region overcoming the doctrine.'

The above quotations - and there could have been many, many more - prove one thing above everything else: the deep-seated nature of the myth. The analysis of this reality will lead us in the opposite direction. For many centuries the Mediterranean has been periodically devastated, fragmented, as a result of conflicts. Life in the Mediterranean is broken into several pieces; every part pushes towards the outside while the outside, in turn, determines it. It is neither a geographical unit nor a cultural area any more.

THE SEAS

The Mediterranean is an enclosed sea. Is it therefore a calm one? Certainly not. It is a sea broken up into a number of basins; a sea agitated by surface currents and by storms; an exploited and an ill-treated sea. It is, above everything else, an open sea.

It is open, basically, because its depends on the Atlantic Ocean. Without the Straits of Gibraltar and the enormous mass of water which has been flowing into the Mediterranean for ages, the sea itself would cease to exist. The great rate of evaporation which takes place in the sea would lower its level by one metre a year; the rains and the rivers only supply it with an average of 900 km^3 of water annually. The Atlantic supplies it with 38,000.

With an area of 3 million km^2, the Mediterranean is only a small fraction of the expanse of waters on the planet: $^1/_{180}$, this being equal to only $^1/_{35}$ of the Atlantic Ocean. Measured along 35° latitude North, it is 3,800

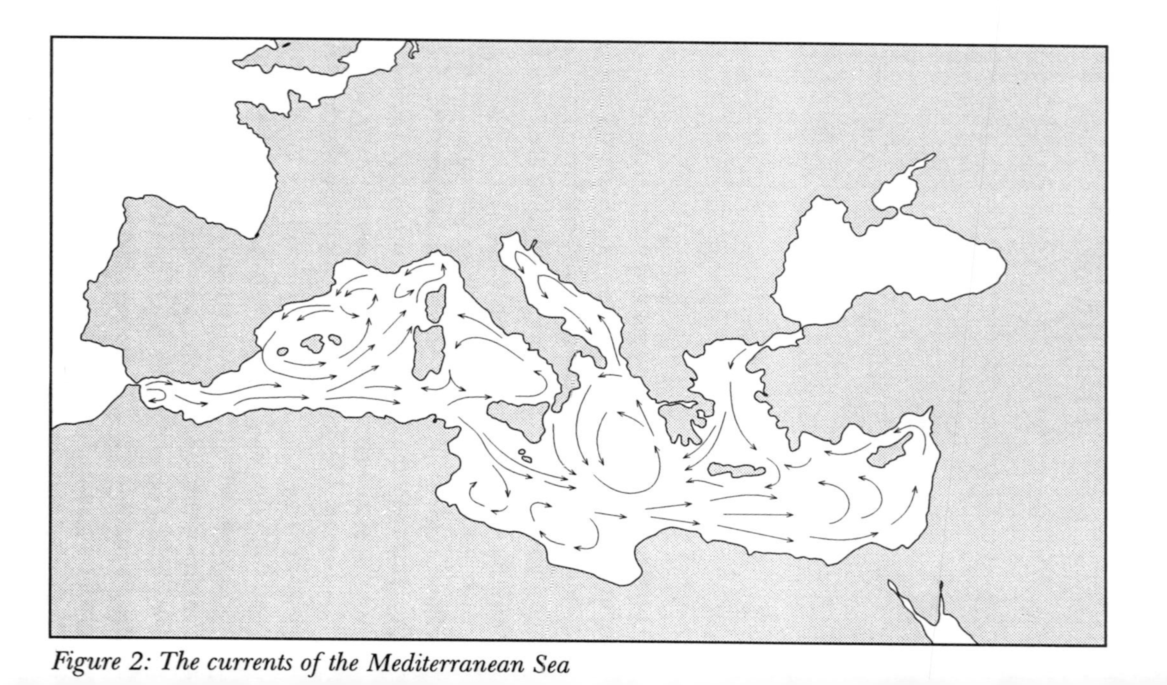

Figure 2: The currents of the Mediterranean Sea

km long. At its widest, between the gulf of Genoa and Tunisia, it measures 800 km. Its narrowest part lies between Sicily and Tunisia where it is only 138 km wide. There, too, a depth of 135 metres divides the sea into two clearly-identifiable basins: the western and the eastern ones. Fortunately, the Mediterranean is a deep sea, an average depth of 1,500 metres, and in some trenches, such as south of Greece, it is over 4,000 metres deep. The volume of the water thus compensates for its small size.

The basin of the Western Mediterranean was formed by a process of subsidence and is surrounded by ancient geological structures whose tectonic nature can be seen in the volcanic activity of the region (Mount Etna, Mount Vesuvius, Stromboli). The Eastern Mediterranean is of wider dimensions, since the subsidence has created seas with their own particular dynamics: the Adriatic and the Aegean Seas. The Straits of Gibraltar are 12 km wide, 58 km long, and 935 metres deep at their deepest.

The State of the Waters

The Mediterranean, as navigators are well aware of, is not a calm sea by any stretch of the imagination. It is kept in a state of agitation by currents which are far from negligible. The surface current which enters through the Straits of Gibraltar brings about an extensive anti-clockwise motion in the western basin that makes itself felt even in the eastern one where it picks up speed as a result of the fast currents of the Dardanelles Straits which flow south from the Black Sea. Although the movement is not very strong, the continuous undulations, together with short and deep waves, give rise to conditions of bad weather which, as a result of the extremely stormy seas they cause, are feared with good reason. Finally the movement of the tides, contrary to a widely-held supposition, is a normal one, although it is very weak since it almost never exceeds 20-30 cm. It can, however, reach up to 2 metres in some particular circumstances, such as in the gulf of Gabes.

The sea is generally calm throughout the summer months. However, navigation becomes very dangerous around the summer solstice, as numerous shipwrecks bear witness. Divers regularly discover the remains of magnificent vessels still carrying their precious cargoes of columns, marble statues, and thousands of amphorae used for the transport of wine and oil on the bottom.

The Mediterranean Sea is warm, salty, and blue: one cannot gainsay the tourist brochures. A warm sea? Its latitude and its particular geologic position, enclosed as it is between mountains and deserts, explain why in February, the coldest month, the surface temperature still reaches 12° off Provence and 17° in its eastern basin. In August the temperature generally reaches 22° around the French coast, 25° off Andalusia, and 27° off Egypt. A salty sea? Since the inflow from the Ocean does not make up for the losses in evaporation, its salinity increases the farther one is from the Straits of Gibraltar. The salinity varies from 3.6 to 3.9 parts per thousand, while that of the Atlantic is 3.5. A blue sea? There can hardly be any doubts about its deep intense colour; it is not without reason that the French call it *la Grande Bleue*. The reflections of a blue which is often dense and without shades, are accentuated by the transparency and the limpidity of the water. Moreover, the water is poor in organic content and in plankton. It is, to put it briefly, a poor sea!

Fishing

The natural environment does not greatly favour fishing. The continental shelf is generally rather narrow and there are no great currents to bring the minerals and the organic deposits of the bottom up to the surface. Because of this the waters 'exhaust' themselves and are renewed very slowly. As a result there is not that great concentration of fish such as one encounters in the oceans; on the other hand, the varieties of species are worthy of note and are accentuated by the division of the sea into its various basins. Fishing in the Mediterranean can only be carried out by small boats with limited holds and which remain at sea for short periods of time. Still the fishermen's efforts and the freshness of their product have created a luxury market. If, in the past, fishing was - as it still is - an insufficient source of food, it still remains an economic resource that cannot be ignored on both an individual and on a family scale.

But the spaces for the fishermen of the

Mediterranean are actually being continuously restricted in a most worrying manner. The coastal waters are polluted and are slowly losing their fauna while the construction projects and ports that tourism demand along the shores contribute further to suffocate the traditional fishing activity. At the same time such resources are being exploited to the full, and sometimes even beyond that, since fishing is disorganized and badly regulated. Some species of fish, which in the past used to be truly 'miraculous', such as tuna, are slowly being wiped out. Some particularly generous regions, such as the Bosphorus, provide smaller catches than they used to. Italian and Spanish trawlers go out to fish into the Atlantic, while large-scale fishing fleets complete with mother boats gives rise to continuous arguments, like those between the Libyans and the Japanese. All in all, the consumption of fish in the Mediterranean regions today depends on imports.

Navigation

For the peoples of antiquity, the Mediterranean appeared as an immense sea. At first navigation only consisted of coast-hugging and whole centuries had to pass before some courageous navigator dared sail across it. The technological evolution which brought about the building of boats and the evolutions in politics, that is in conquests, and in economic conditions – basically in large-scale commercial exchanges – slowly turned the sea into a place of ever-growing traffic. Thus even space-time changed. A sailing boat could take up two months to cross from Gibraltar to Istanbul, depending on the caprices of the wind. Today we can make a crossing in any weather conditions in a few hours, unless one takes an aeroplane in which case a few minutes would suffice. The Mediterranean Sea is a lake ... but not completely; it is indeed a lake through which ships pass rather than sail between its various ports.

The Mediterranean is an area that gives rise to new enterprises. It was a Mediterranean man, born in Venice in the twelfth century, who discovered the route to China, while it was another Mediterranean man, born in Genoa in the fifteenth century who first crossed the Atlantic. Long-distance trading enriched and brought about the flourishing of the great Mediterranean ports: Venice, at the centre of a vast commercial network from the fourteenth to the sixteenth century, was the richest city of the Mediterranean and of the entire European continent. The decline started when the English and the Dutch moved the most important commercial flows to the west. Then, starting from the seventeenth century, they made use of external outlets for Mediterranean trade, taking over Venice's space. 'Not even the building of the Suez Canal would restore fully the former prosperity and, above all, the pre-eminence of the Mediterranean' (Braudel, 1985).

At any one time at least 2,000 commercial vessels with a gross tonnage of at least 100 tonnes, including 300 tankers, are sailing across the Mediterranean. More than 200,000 crossings are estimated to take place every year. After the English Channel, this is the area with the heaviest sea-traffic in the world. Outside every major port, near the straits, and near the Suez Canal the crowding is easy to see. At Gibraltar the territorial waters overlap and an international treaty had to be signed to regulate the use of the sea-lanes. Still, navigation on both a national level – which cannot be ignored in the case of countries such as Italy, Greece, or Turkey – and on an international one, represents in all only 20 per cent of the traffic which is essentially trans-Mediterranean, whether it brings these ports in contact with the rest of the world or whether it just makes use of the sea as a passage without affecting any calls on Mediterranean ports. About 70,000 vessels sail through the Straits of Gibraltar every year, not taking into account the ferry-boats, the fishing boats, and pleasure craft, and not to mention the boats that carry illegal immigrants or the powerboats that smuggle hashish and cigarettes to Europe. The Suez Canal, on the other hand, was used by more than 16,000 ships in 1992; its widening to 23.5 metres and its deepening to 56 feet now enable ships of a tonnage up to 150,000 tonnes to pass through. Further work – at the moment not judged economically feasible – will make it navigable for ships of 270,000 tonnes. The competition being provided by the Suez-Mediterranean oil pipeline, SUMED, risks the postponing of this latter project to the Greek calends.

The fleets which sail through the Mediterranean

belong to most of the world's nations and are of the most various types: commercial freighters, tankers, bulk-carriers, and tramp ships in accordance with the laws of maritime economy. The Greek merchant navy merits a special mention for it represents more than half of the carrying capacity of the countries around the Mediterranean coast. Petroleum products constitute 50 per cent of the cargo carried. It is estimated that the transport of petroleum products across the Mediterranean, consisting of loading, unloading, and transhipment, make up at least 20 per cent of the world-wide transportation of the commodity, even though the sea's surface area is just 0.7 per cent of that of the planet's seas. The ports used for loading and unloading by the tankers, in addition to constituting a closely-knit network, are symmetrically distributed on both the western and the eastern shores. Methane gas, of which Algeria and Libya are important exporters, has to be liquefied to be carried in ships fitted out for the purpose with sophisticated and costly equipment. Underwater pipelines for the carrying of the gas are a far simpler means of transportation since they are simply an extension of terrestrial ones. That which joins Algeria to Italy by way of Tunisia, the Straits of Messina, and the Sicily channel, has succeeded in gaining a good share of the market; similarly that which unites the Maghreb and Europe through Gibraltar will also claim a good part of this traffic.

Maritime traffic in the Mediterranean is not limited to petroleum products. Considerable tonnage criss-crosses the sea; raw iron ore and coal from distant lands supply the steelworks built on the coast at Fos or at Taranto. The transport of cereals, on the other hand, makes up a noteworthy north-south traffic since Algeria and Egypt have both become considerable importers.

The use of pleasure craft, moreover, has today assumed a primary role, both from the economic point of view and from that of the utilization of large coastal areas. It has been calculated that more than 100,000 persons spend their summer holidays aboard a boat. Marinas for such craft, constructed in the old fishing ports or, more frequently, simply built from scratch, are considered as essential for 'complete' tourist bases. Maritime resorts are complexes in which tourist ports form a fundamental element, in the Mediterranean, on average, they have 400 mooring places and cover tens of hectares of land. But many of these maritime resorts exceed that number considerably, sometimes reaching 1,000 berths or more. The European coasts, obviously, are the better equipped since they are more accessible for this particular type of client and because they benefit from a more comfortable environment. Italy has 445 ports available for navigation by tourists, some of which form part of very large seaside resorts: Riva di Traiano not far from Rome, Rapallo on the Ligurian coast, and Porto Cervo in Sardinia. Even Spain, with Ampurias Brava on the Costa Brava and Marbella on the Costa del Sol, is in the forefront, rivalling France which has built resorts at Port Camargue and Port Leucate in the Languedoc and Port Grimaud. The islands which are not situated far from the coasts also benefit from such intensive business: the Balearic islands, Sicily and Sardinia, Corsica, the Cyclades, and the Sporades. To the south, the Sahel in Tunisia is updating its facilities at a febrile rate. And everywhere the tourist ports and the hotel complexes that accompany them are subject to lively speculation, mostly by international capital which demands an immediate return, although often dubious, in the absence of public funding.

The Ports

For ages, therefore, the ports have constituted important tesserae of the Mediterranean economy. Certainly there is no shortage of natural sites, although their small dimensions resulted in them being almost ignored or else enlarged artificially. At Tyre and later at Alexandria, the people of antiquity showed their ability in constructing breakwaters using heavy blocks of stones, while artificial basins were built at Carthage and Ostia making use of alluvial matter to build, thanks to a sort of hydraulic cement, the banks on the shores. Also ancient times gave start to particular specializations, like the functional one between shipbuilding and ship-maintenance; the professional one between ship-owners and ship-captains; and those of a commercial nature. Piraeus was an emporium where, during fairs, it was possible to find commercial samples and where the network of redistribution was perfectly organized.

Today petroleum dominates harbour activities in the Mediterranean and there are fifty-eight ports in the entire basin which can be used for this trade, sixty refineries, and a hundred or so oil-fed power stations.

The big ports are actually great cities. Even if harbour activity today only plays a secondary role and cannot rival that of north-western Europe, it is still vital for the Mediterranean economy. Marseilles, with its annexes at Fos and the dockyards at Berre, is the third European port, especially as regards petroleum-connected activities: its import, refining, and export by means of the South-European pipeline. In the other sectors, which should assure the possibilities of expansion and diversification, Marseilles suffers, according to an expert, 'from a too strict dependence on the Ocean'. The port, together with the city, is a victim of the slowing down of public and private investment and of the adoption of technological innovations that reduce employment, while suffering from an endemic inability to resolve social conflicts. Similar problems face Italian ports, open as they are to the Atlantic and the seas beyond the Suez Canal; two hegemonous zones which subdivide the access and the commercial traffic towards the industrial regions of central Europe. The Ligurian Sea has harbour complexes (Genoa, Savona, La Spezia, and Leghorn) which are highly dependent on the importation of petroleum, but which have learned to specialize - by providing platforms and access roads in their hinterland - in the transportation of containers. The north Adriatic, on its part, with Ravenna, Venice-

Marghera, Montefalcone, and Trieste, benefiting for the time being from the closure of the port of Rijeka (Fiume) in the former Yugoslavia, caters for the commercial demands of the Po basin. Naples, naturally, remains an important port for many reasons, and not just for southern Italy: industry, tourism, culture, and the armed forces all draw benefit from its siting and its unique location.

In Spain, although the Mediterranean ports do not figure in the forefront in terms of tonnage, the diversification of sea traffic still allows them to play a very important economic role. Even Tarragona, which specializes in hydrocarbons and petrochemistry, and Algeciras carry considerable amounts of sundry types of cargoes. Valencia has less business, although this is well-balanced since it is regularly used by 280 shipping lines, which account for three-quarters of the movements. Barcelona, finally, is not a major port by European standards but the processes of modernization it has undergone - especially in the handling of containers - are starting to bear fruit. Still, Barcelona is a great metropolis which dominates and serves an area of production and consumption strictly dependent on it, and which has the ambition to become - as it did with the organization of the Olympic Games in 1992 - the leading city of the Mediterranean.

On the sea's southern shores, Arzew is important for the exportation of petroleum while Alexandria is a major port through which pass the vital supplies for Egypt. On Turkey's Aegean shores, most of the activity is concentrated in Izmit, followed by Izmir (Smyrna) and Istanbul.

Greece, finally, which is a case apart, has 128 large and small ports, which are mostly dependent upon coastal trade and cruise liners. Through Piraeus, and, on an even greater scale, Thessaloniki pass the country's heavy cargoes, hydrocarbons in the first place and supplies coming from all over the world (aluminium, for example, is imported all the way from Australia). Greece's specialization, anyway, is something else. The great ship-owners of this small country together own the world's second merchant fleet – flying both Greece's flag and that of other nations – and dominate the world 'tramping' market, with their ships sailing from one harbour to another in search of cargoes.

Pollution

Obviously, the devastation caused by pollution is mostly visible close to the shores, starting with the large blobs of black oil that stick to the rocks or lie hidden in the sand. To these one must add the solid wastes that sink to the bottom or that float on the surface. Even more dangerous, because they cannot be seen, are the bacteria which multiply in the outfalls, not to mention the chemical products and the residues of metals such as lead and mercury.

To assess such pollution serenely and objectively, especially if one keeps in mind the area of the sea involved, is not easy. The dramatic appeals for help for the 'murdered Mediterranean' and the scathing declarations of certain ecology gurus which declare the sea to be 'finished' (1987) contribute on their part ... to pollute the scientific truth.

Certain facts are, however, beyond discussion. The hydrocarbons are rightly held much to blame. The experts rejoiced when the amount of oil discharged into the sea decreased from 800,000 to 650,000 tons – even if that were actually true – over ten years, and they say that two-thirds of it evaporates without causing any environmental damage. Fortunately, accidents are rare, but ballastage, which has been banned by international treaties for many years, is still a common occurrence, mainly because of the scarcity of proper port equipment.

The total weight of refuse of all types from the entire continent is estimated to amount to around 10,000,000 tons, that is fifteen times that of hydrocarbons. It consists of water from domestic sources (70 per cent of which finishes directly in the sea); untreated industrial waste; and, most of all, of the effluents often brought from far away by the great rivers (the Po is the major culprit in this regard). To these one must add the noxious effects from the large increase of tourist ports and other structures of concrete built in the sea. Such constructions are, beyond doubt, responsible for the disappearance of the poseidonia meadows even in the wildest localities. These expanses of seaweed provide natural shelter and nutriment for numerous species of fish.

Bacterial and viral pollution, invisible to the naked eye, which often makes swimming dangerous –

without, however, discouraging the tourists – has been controlled along important stretches of the seashore, primarily in the western Mediterranean. But chemical pollution and the effects of atmospheric pollution, which are responsible for the discharge into the sea of small particles that are not normally taken into consideration, have probably even worse consequences; in the long run, in ways difficult to monitor, they can cause catastrophic mutations in the biological chain, among both the flora and the fauna. It is worth remembering, in this respect, that in spite of the technological advances in the treatment of effluent, it is the northern countries which are mostly responsible for this pollution. France, Italy, and Spain are between them responsible for about 60 to 70 per cent of it. It is this fact that gives rise to the well-justified claims by the southern countries for compensation in the various international conferences; up till now, however, the accused have always limited their contribution to suggesting the buying of proper treatment plants!

Eutrophication is the most spectacular result of pollution and it affects whole expanses of the sea and the lagoons. The discharge of material from agriculture, industry, and urban refuse into the sea blocks the oxygenation of the water, causing the death of fish and the proliferation of particular algae. This phenomenon is well-known, for example, in the gulfs of Saron (Athens) and of Tunis and in the Venetian lagoon where, in some years, this invasion by the algae brought in its wake an invasion of insects and the liberation of hydrogen sulphides and necessitated the use of extensive equipment in an effort to save the tourist season. In the north Adriatic eutrophication has brought about, in summer, the appearance of 'red water' as a result of the accumulation of highly toxic phytoplanktons which annoy the tourists and, above all, greatly interfere with the marine ecosystem.

It is not yet possible to estimate fully the consequences of the appearance in the Mediterranean in the 1980s of *Caulerpa taxifolia*, a tropical seaweed which grows about 2 cm a day and which has a tendency to cover the sea bottom - at a depth of between 8 and 50 metres - and then to kill off all the other vegetation. It is truly a 'murderous seaweed', but is it really responsible for an 'ecological catastrophe' as some experts proclaim? Nature, probably, has not said the last word.

In this, as in other cases and processes taking place, once the danger has been recognized, the overall situation of the Mediterranean has to be considered objectively. Have the scientists who, in the Barcelona conference of 1995, presented their action plans for the Mediterranean, the famous Blue Plan launched in 1995, not established perhaps that 'the offshore waters and the sediments are of a relatively-accessible quality, comparable that of the high seas in the oceans'?

Surrounded by twenty states, all of different sizes, levels of development, and political status, the Mediterranean is an odd sea. If one thinks of the total area of the sea and the protection it needs, it has to be seen as an indivisible whole. It is not surprising, therefore, that today scientists see the need for international co-operation as evidence for its unity. In return, it is surprising that such evidence has only pushed politicians to utter smooth words, to propose timorous projects, and to decide on collective interventions only on a very limited scale.

The International Commission for the Scientific Exploration of the Mediterranean was set up in Monaco in 1910. However, it was only in 1975 that the first Mediterranean convention was signed by all the states on its shores: Spain, France, Italy, Albania, Yugoslavia, Greece, Turkey, Syria, Lebanon, Israel, Egypt, Libya, the three Maghreb states, Monaco, Malta, and Cyprus. The European Economic Community also signed the convention.

As a 'moral person' the Mediterranean, therefore, has a special fund paid for by the various nations in proportion to their gross national product. The executive of this organization, which is known as the 'Action Plan for the Mediterranean' (APM) has its seat in Athens from where it administers the project which has six different thematic centres. The most important is MEDPOL, which co-ordinates research and the collection of data about pollution to establish relevant regulations. The Malta centre is concerned with the prevention of these risks relative to the transportation of hydrocarbons and of chemical products. The other centres are concerned with the protection of natural and historic sites, development, and research. This latter centre, situated in Sophia-Antipolis, in the hinterland of Cannes, is known as 'Regional Action Centre for the Blue Plan', and is entrusted with pointing out, by means of 'a global and systematic approach of the Mediterranean basin considered as a whole', both short- and long-term evolutionary tendencies and critical problems.

Anyway, the application of the Barcelona convention is marking time. It was necessary to wait ten years until, during the Genoa conference of 1985, a number of precise and urgent objectives were set, including the construction of stations for the transfer of oil on the high seas and the elimination of petroleum residues in the ports, the building of waste water treatment plans for the cities, the substantial reduction of industrial and atmospheric pollution, the protection of threatened marine species ... Objectives or wishes?

At the same time, some state intervention projects have helped to involve new partners, such as the World Bank and the European Investment Bank, which have made possible the setting up of complementary projects. In 1990 these two institutions, on the initiative of the EEC, signed the Nicosia document with most of the seaboard countries and pledged financial assistance for identified projects.

A number of scientists and politicians would like to

see in this stuttering co-operation the blueprint of a real and actual Mediterranean community. This is not, however, a very realistic position; the Mediterranean is too open to the world and too dominated by the North to make possible the creation of a collective identity. Unless, as Serge Antoine, the promoter of the Blue Plan, writes, this Mediterranean identity is not defined in relation to the recognition of its 'multiple belonging'.

The Spaces

Participating in conferences or in co-operation programmes for the protection of the marine environment does not pose any problems, if not of a diplomatic character, for the countries of the Mediterranean seaboard. Matters, however, assume a very different dimension when questions of sovereignty over territorial waters are at stake. International conventions are drawn up in such a way that, if applied to the narrow, compartmentalized, and almost enclosed Mediterranean, they become inevitable sources of almost insoluble conflicts. The establishment of jurisdictional limits and the territorial rights by the various countries give rise to interminable confrontations, fortunately not of a belligerent nature. Mostly they concern problems connected with territorial waters, straits, the continental shelf, and some exclusive economic zones.

In principle, territorial limits extend to 12 nautical

miles, and this rule also applies to the Mediterranean since the signing of the Montego Bay Convention of 1982 which establishes a straight base line. In this area, the coastal state is given control of customs, health, fiscal and immigration matters, together with full rights as regards the protection of the underwater heritage. Moreover, the state is allowed to exploit all of marine resources therein. As a result, the various states exert sovereignty over one-third of the total surface area of the sea.

Navigation through straits is controlled by the Montreux Convention of 1936. On the strength of this convention, Turkey, the guardian of the Dardanelles, may deny passage to warships (the Soviet submarines used to come from the Atlantic). By extending the limits of territorial waters, the Montego Bay Convention has artificially increased the number of straits in addition to the natural ones: there are fifteen of international relevance. Only the Otranto Channel and those of Malta and Sicily, which still contain a tract of open sea, are subject to rules of common law. The others are subject to different jurisdictional situations although the states concerned can neither suspend nor impede inoffensive traffic.

The continental shelf is the main source of conflicts. The Montego Bay Convention gave a rather loose definition: 'The natural prolongation of the terrestrial territory until the outermost limit of the continental margin or up to 200 nautical miles from the base lines ... of the territorial waters.' Now, in the Mediterranean there is no continental shelf which stretches out to 200 miles, since the greatest distance between two shores never reaches 400 miles. All problems, therefore, have to be faced case by case. Some bilateral agreements have been signed, but recent events have been marked by serious controversies between Greece and Turkey and between Libya and Tunisia. These controversies, following the breakdown of long and often dramatic negotiations, have been taken to the International Court of Justice, and the desire to find equitable solutions has led to the proposing of specific solutions.

Finally, the notion of an 'exclusive economic zone', as proposed by the Montego Bay Convention, in principle allows states to exercise a general sovereignty of their own within the 200 miles limit. The prospects of conflicts such an initiative would raise because of the particular configuration of the maritime spaces have in practice stopped the coastal states from availing themselves of these opportunities. The fact remains that, according to some observers, in the Mediterranean 'the high seas are in a moratorium'.

THE ENVIRONMENT

Considered as an archetype, the environment of the Mediterranean reveals the limits and the resources of the entire basin: its geographical position, its fragmentation, the protective belt of mountains and deserts, the antiquity of the first human settlements, and the present over-population. From these primary characteristics and other secondary ones, both on a regional and on a local scale, there emerges a symbolic landscape with a rarefied presence but which can nevertheless be encountered everywhere.

The Landscape

The landscape is characterized by the frailty of the vegetation and by the fragmentation of the natural units. Over the centuries the great forests have become even rarer and vast areas are today covered by

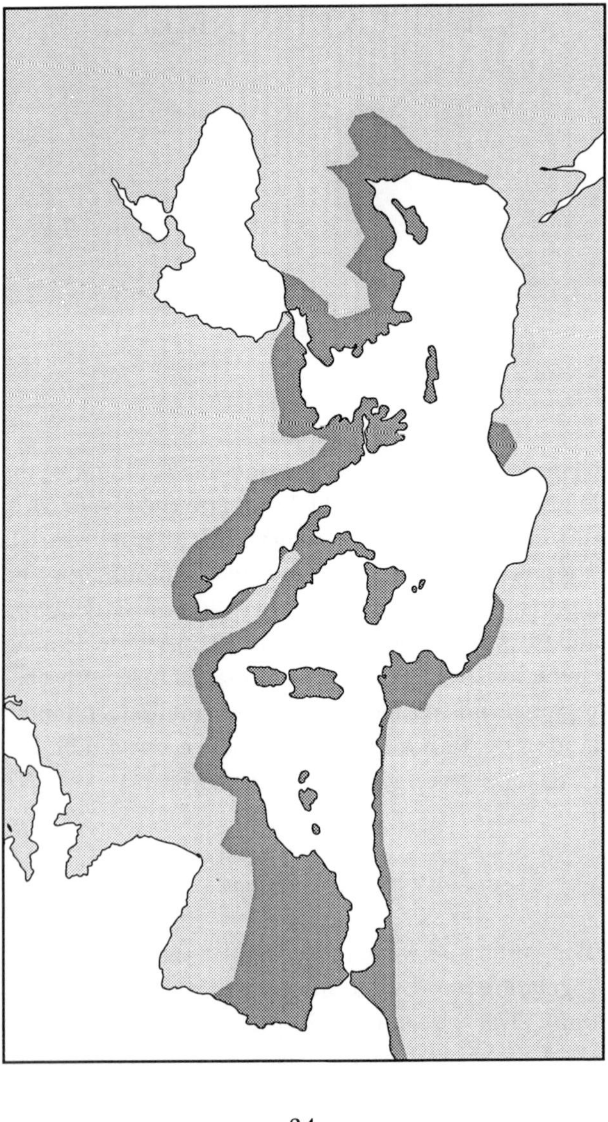

Figure 3: The diffusion of the olive tree

degraded formations or thinly-wooded plains. In the impoverished soil on the sides of mountains and steep hills, one can see rocky outcrops scoured by the waters. The watercourses, often of fairly large dimensions but dry for most of the year, lead to small plains and to the coast. Everywhere there is the imprint of man that bestows these landscapes with their originality. From the most ancient agrarian civilizations comes the classic division into three distinct parts but which are co-ordinated within the area a community occupies. According to the terminology of the ancient Romans, the regularly cultivated *ager* is harrowed and sowed and in general it occupies either a plain or the more favourable slopes. The *saltus* is not cultivated regularly or continuously but is used for the pasture of livestock and its vegetation forms notable expanses of bushes: maquis on siliceous soils and garigues on lime ones. The *silva*, finally, on the borders of the land and on mountaintops often consists of a stunted forest as a result of the excessive exploitation it has had to withstand, although the undergrowth is often quite luxuriant.

Together with the legendary olive tree, the holm oak, and lavender, there are two other trustworthy indicators of the Mediterranean environment, whose image is associated with exotic plants which are encountered in all landscapes and in agricultural productions to such an extent that they can be considered as having been 'naturalized'. About one-half of the Mediterranean flora, taking into account all the species and varieties, is endemic. However, the cypress comes from Persia, the plane tree from Asia Minor, and the eucalyptus and the mimosa from Australia. The agave, which holds the terraced soil, and the prickly pear come from the New World. The tomato, that almost symbolic agricultural product, came from Peru, before it was generally diffused thanks to greenhouse cultivation, while citrus trees came from China. This is one more argument in favour of the idea of a great, open, and dependent Mediterranean.

In spite of appearances, the forest represents the apex of the vegetation of the Mediterranean basin. Still it has suffered so much as a result of the breaking up of the soil, fires, and the exploitation of timber for ship-building and for heating, that today there are only a few scraps that provide a poor testimony. It has been estimated that no more than five per cent of the primeval forest survives in the Mediterranean, and that consists mostly of oaks and conifers which dominate an underwood of evergreen trees. The holm oak, which is the most diffused species, flourishes with the lentisk, the red oak, the juniper, or with heather or rockrose. It is also possible to find some survivors of the cork oak on coastal massifs of crystalline rock. The various species of pines are distributed in accordance with the different local contexts: the cluster pine grows on acid soils, together with heather and ferns; the umbrella pine flourishes along the coast; the Aleppo pine is a common feature on lime soil garigues. Planted and tended by man, there are other forests which grow on the middle altitudes, like the chestnut forests in Corsica. On the higher altitudes, in the north, one can still encounter beech trees, Scotch pines, and fir trees, and Lebanese cedars in the south and the east.

The maquis is dotted with oak trees, the last survivors of the primeval forest, and which, together with heath and broom, is held to be impossible to destroy completely. In fact it is so little dense that goats and sheep can find what to graze upon while hunters and hunted are able to weave numberless paths all through it. The garigue consists of low and spiny bushes and is relatively open; it does not cover the rocky ground completely, outcrops of which can be seen here and there. In southern France it is characterized by red oaks, in Spain by the myrtle and the lentisk, and in North Africa again by the lentisk and by the wild olive tree.

Garigue, maquis, and forest suffer very greatly from the danger of fire. Fires, often devastating, destroy thousands of hectares of vegetation every summer but nature seems to looks after its wounds since the damage is made good in a few years, in spite of the indifference of man. Actually these fires contribute, more than anything else, to the irreversible degradation of the natural environment. Fire needs an abundant maquis to spread: first of all, a bed of dry leaves and then a low-lying wood. The maquis itself has spread as a result of the gradual abandonment of the traditional farming and pasturing system. The time has come to solve the problem of the relative costs of the work necessary to remove the brushwood (either by means of machinery or by devoting areas to pasture) and to fight the fires. The prevention of fires is a rational objective: one has to consider seriously the risks connected with urbanization, the uncontrolled development of tourism, and the remaining activity of large-scale stock-raising for which it is indispensable to keep open spaces and to favour the growth of grass ... by using fire.

Typical steppe vegetation covers the drier areas. Although this was part of the classical Mediterranean landscape, it still occupies vast areas in the south and in the east: the plateaux of the Maghreb; those regions of Libya and Egypt between the sea and the desert; the east; Anatolia; and also the central regions of the Iberian peninsula. Esparto grass grows on its stony soil, while artemisia prospers in clayey conditions.

In the marshlands and the river deltas little has survived of the original landscape. Such zones, distinguished by marshes and lagoons and always hostile to permanent human settlement, have been conquered over and over again. In recent years they have been totally transformed by rapid colonization. The plains of Lazio and the Campagna have witnessed the completion of a great project of draining of marshes first started in Roman times. Even the *marismas* of the Guadalquivir have been reclaimed, just like the marshlands of Macedonia and the Mitidja in Algeria. All in all the marshlands, which measured 3,000,000 hectares in Roman times, today extend less than 200,000 hectares ... to the great dismay of ecologists.

In order to meet the demand for food, the steppes and the low plains in the south and the east of the Mediterranean basin have been used for both dry and irrigated cultivation, with a notable effect on the natural environment. The areas used for the dry cultivation of wheat in the steppes has increased at least threefold since the Fifties, with the result that the tendency towards desertification has been speeded up in North Africa and the Middle East owing to wind erosion. On the other hand, intensive irrigation in recently-reclaimed plains has increased salinity. It has been estimated that in such conditions, about 30 per cent of the Egyptian agricultural potential cannot be utilized.

The Climate

Geographically, the Mediterranean lies on the northern edge of sub-tropical high-pressure currents. In summer, the northwards drift of the Azores anticyclone pushes back the unsettled weather that comes from the middle latitudes in the west and brings the dryness the desert. In winter, its southwards movement leaves an open door for air currents from the North Pole: the polar front and, more frequently, the Mediterranean and Saharan one cause showers. The alternation between the dry and hot and the rainy and cold seasons, the result of the geographic opposition of the climatic conditions, is a characteristic of the Mediterranean. First of all because the north and the south are so different at their respective extremities: in the north the mistral and the bora cause a fall in the temperature, while in the south the scirocco and the *khamsin* make it rise drastically. Moreover, although the mountainous reliefs around the basin serve almost everywhere to protect it from climatic excesses, they are responsible for producing a marked difference between the Mediterranean continental climate and the Mediterranean littoral climate, the latter typical of the islands and a strip of coast a few kilometres deep. Finally, the compartmentilization of the reliefs creates the right conditions for stabilizing local climates or microclimates.

The average and the extreme temperatures bear witness to the existence of these diversities. The warmth of the sea is the reason why the average January temperature is 12.2° in Malaga, 12.3° in Naples, and 12.9° in Tripoli. Inland, the average January temperatures are not much lower: 7.3° in Granada, 6.5° in Potenza, and 10.5° in Jerusalem. However, it is waves of cold weather, carried by currents of air from the polar regions that mark the winter conditions. Freezing conditions can be experienced in Barcelona and in Palermo, and even more commonly in Granada and Florence. Sometimes ice has been known to form in the gulf of Thessaloniki (and even in the harbour of Marseilles). Certain cold spells, like that of 1956, for example, affected the olive trees of Provence, the orange groves of Andalusia, and the vineyards of the Languedoc. In summer, temperatures are not less marked, although the average July temperature is about 25° in Alicante, Candia, and Tel Aviv, although it can reach 30.7° in Tripoli and 30.9° in Jericho. Heat waves, however, are much feared: they can even reach and surpass 45° in Athens, and also in Palermo, Kairouan, and Cairo.

More than the temperatures, it is the rainfall that gives rise to the greatest contrasts. Rainfall, at least in the northern regions of the Mediterranean basin, is quite plentiful: the annual average is about 400 mm in Athens, 500 in Barcelona, and 600 in Nice, although it reaches 1300 in Genoa and 1200 in Corfu. The southern and eastern regions are drier, without, however, being arid along the coast. At Algiers the average rainfall is 550 mm, 400 in Tel Aviv, 200 in Tripoli, and 150 in Alexandria. Still this rainfall is seasonal and concentrated in the winter months and over a small number of days, in general from 40 to 60 days. Still, the difference in rainfall from one year to another can be quite considerable. In the space of ten years, the annual rainfall in Marseilles registered differences from less than 400 mm to more than 800. The duration of the dry season, therefore, varies greatly, depending on the locality but, above all, it varies from year to year and lasts from 5 to 8 months.

All in all, the tourist has little to fear from the weather. The Mediterranean continues to enjoy long periods of good weather, while bad weather only comes in spells and can change very quickly. The coasts enjoy more than 2,500 hours of sunshine in the west and 3,500 in the south and the east.

The Mountain Reliefs

The collision of the African and European continental plates in the second half of the Tertiary Age – just some tens of millions years ago – caused a fault in the planet which gave rise to the Mediterranean Sea and also raised the mountains that encircle it. The permanent seismic nature and the volcanic phenomena bear witness to this on-going tectonic activity. Neither are flatlands absent – either as high plateaux as in the Spanish *meseta* or in Anatolia – or along the southern coasts from Tunisia to Arabia. It is, however, sharp slopes, broken into small units at the sides of high mountain chains, that predominate.

The western Mediterranean is completely surrounded by mountains: from the Alps to the Apennines, from the Pyrenees to the Sierra Nevada, and from the Atlas mountains to the mountain chains along the Maghreb coast. The northern regions of the eastern basin, on the other hand, are surrounded by the Dinaric Alps and by the mountain chains of the Balkans, the Taurus, and Lebanon. Because of the large reserve of water it holds, this mountainous chain is not only the *raison d'être* of life in the foothills, the hills, and the plains but also the cause of their fragility. And this the farther one gets away from the more or less isolated high peaks, which are an integral part of the Mediterranean scenery: the Mulhacén above Granada, the Taigetus above Sparta, Mount Parnassus and Mount Olympus, the Mercantour (Argentera) and the Canigou, without forgetting Mount Vesuvius and Mount Etna.

At a lower level, the alpine orogenesis is also responsible for the fragmentation and the structural diversities which give rise to the rocky outcrops which in turn assume many different forms. Ancient material, pushed from below, surfaces in crystalline massifs which form unique structures with their typical vegetation: the Mauri and the Esterel, the Cabilia, Sardinia, and Calabria. The sedimentary matter, flat or undulating, and always traversed by fissures, is mostly of limestone origin, which makes the effects of aridity worse.

The common factor of these environmental patterns is a certain fragility, owing to the ongoing tectonic activity, to climatic conditions, and to the action of man. One can almost state that, paradoxically, nature is traditionally unstable. How then can one explain the fact that many ancient buildings are still partially standing and that so many remains of the past have not been destroyed by erosion or buried underneath alluvial matter? The answer must be surely sought in the exceptional character of the paroxysmic present that has succeeded the long phases of relative equilibrium which had been guaranteed by the protection afforded by the forest. Demographic pressure, together with that exerted by the search for pastures, is therefore the main agent of the derangement that characterizes nature at present.

Local erosion, that which is noticed by visitors, is actually caused by deforestation, by the cultivation of hillsides that are too steep, and by the building of roads. The absence of trees allows violent rainstorms to create streams that break the soil and carry it away. On the other hand, in the early Sixties agronomists established that a third of the cultivated area in the south of Italy has a gradient of 15 and even 30 per cent. The result can be seen in landslides which are actually the results of the imbalances produced by erosion. It is calculated that from 2 to 3,000 landslides of a certain magnitude take place every year in Italy. In addition to the landslides caused by unstable formations like clay and marl, there is also the erosion caused by barrancas, calanques, and by the 'bad lands'. The scouring of the ground creates a landscape that makes one think that 'the earth is dead, and that only its skeleton has been left exposed to the sun, washed and whitened by the rains' (C. Levi). The net result is that the swollen rivers and the streams carry away impressive amounts of soil. In the Bradano and in Basilicata, suspended solids have been measured at 70 kg for every cubic metre. Attempts have also been made to calculate the average depth of this annual ablation: it is calculated that on the Adriatic side of the Apennines this erosion amounts to more than one millimetre.

Below the denuded rocks, the vast stretches of rivers, torrents, and *wadis* – the *ramblas* and the streams – do not serve to drain the water, unless they are fed from the mountains, except for a few days every year. On such occasions, the solid wastes they carry to the lower areas fill the mouths and give rise to deltas. The inner plains and those along the shores would benefit, if such alluvial matter did not bring about the formation of lagoons and marshes.

The limestone shelves and slopes, which are often encountered near the shores, resist erosion since they have already lost their soil owing to deforestation. The white rocks appear almost completely denuded and end up by becoming a typical element of the Mediterranean landscape. In their fissures, cracks, and holes, the red soil gathers, testifying to processes of decalcification that have been going on since ancient times.

Tectonic instability, which is the cause of earthquakes and volcanic activity, is the result more of the Mediterranean's natural history than of processes of erosion. Anyway it carries grave risks so far as the densely-populated areas are concerned. Volcanic activity is limited to specific places; only the Vesuvius and the Etna are really dangerous. The Aegean island of Santorini is a crater which was shattered by an explosion around 1450 BC. The explosion covered Crete with stones and dust and almost certainly contributed to the destruction of that splendid civilization.

Seismic activity is widespread. According to archaeological evidence, catastrophic earthquakes have been a feature of Mediterranean history since ancient times. They still occur, sometimes more than once in the same locality, but are only remembered when they claim victims; in this last millennium a hundred or so devastating earthquakes have occurred. The one in Lisbon in 1755 caused tens of thousands of victims, like that of Messina in 1904. Nearer our time, the earthquakes in Agadir, Skopje, El Asnam, Campania-Basilicata, and the gulf of Corinth remind us that there is no place which is absolutely safe.

All in all, even in this case of natural violence, the western basin appears less threatened than the eastern one; even nature favours the west.

The coast of the Mediterranean extends to 46,000 km and is determined by the continental relief; its broken nature accounts for the weak development of the continental shelf. In the gulf of Lyons or in that of Gabes, its width corresponds to a relatively low hinterland, or, on the eastern shores of Spain or the central Adriatic, to submerged blocks at the foot of steep slopes or faults. Anyway, as opposed to the small extension of the continental shelf, the stretch of low or built coastline is quite notable and makes up 40 per cent of the entire shoreline. It is the result of the accumulation of material caused by thousands of years of erosion. Even the average river forms a delta of its own, while those of the great rivers – the Ebro, the Rhine, the Po, and the Nile – occupy very extensive areas. Even far away from important watercourses, however, low shores and lagoons that extend for thousands of kilometres are formed by the action of marine currents and by submerged sandbanks. Such coasts are usually marshy, but human intervention has made them cultivable or developed them for the use of tourists. On the other hand, however, the rocky and indented coasts are formed in such a way that the sea bathes the continental relief directly, and their outlines therefore depends strictly on the latter's structure. Faults and transversal folds can favour the formation of rather deep anchorages. In local terms, however, the various configurations of the coastline merge inextricably. In an imaginary landscape, which is still not misleading, the beach lies side by side with rocks, and the discovery of a stretch of sand and pebbles on the sea bottom or a small creek should not surprise us. But is it true, as the experts claim, that 15,000 km of the coastline out of a total of 46,000 have been irremediably eroded or built over with concrete?

In this respect, the islands share the same experiences, but they have a biogeography – and sometimes a total geography – which is heavily influenced by their isolation ... or rather was, since the modern means of communications abolished distances.

Near the coast, the Mediterranean is strewn with islands. Three thousand and three hundred have been counted, but the number depends on the fine sprinkling of small islands and shoals that do not even have a name. Some of them have dimensions, histories, and economies of their own which make them important in their respective regions and some even on a national scale. Of the 2,000 Greek islands, only a few more than 200 are inhabited. Sicily and Sardinia are by far the largest in size, followed by Cyprus, Corsica, Crete, and Majorca. Some smaller ones, like Malta, stand out. Others form archipelagos that stretch out like rosary beads, like the Dalmatian islands, the Kerkennah islands of Tunisia, and, above all, the islands around the Iberian peninsula, the Ionian islands including Corfu and Cephalonia, the Cyclades and the Sporades in the Aegean, and, near Asia Minor, the large islands of Lesbos and Chios in the north and of the Dodecanese, including Rhodes, in the south. The islands, as Matvejević rightly points out, are distinguishable by the image they offer and by the impression they leave: 'Some look as if they can float and sink, others look as if they were at anchor and turned to stone. Certainly they are nothing but incomplete fragments torn from the shore. Others have long abandoned the mainland and, now enjoying independence, are self-contained ...' On some there is a sense of abandon and disorder, 'while on others everything is well-ordered, enough to make one think that it is possible for an ideal order to prevail there'. Islands of dream in the eyes of the modern inhabitants of the mainland; but even places of expiation with their monasteries, or even of exile, with their battlefields and prisons. Daedalus built his labyrinth on the island of Crete, Seneca was exiled in Corsica; the bare island of Guli Otok hosted the dissidents of Tito's regime, and that of Makronesos, the opponents of the Greek junta.

The sorts of dangers that threaten the Mediterranean seashore have already been mentioned since they also directly menace the sea-land interface: urbanization, industrialization, the excessive development of tourism, and pollution of various types. Venice represents a case in point. Following the dramatic incidents of the *acqua alta* of 4 November 1966 and of 23 December 1979, special laws were passed and financial measures were taken to protect the city of the doges. The lagoon, which is connected to the Adriatic by three sea-passages, has a surface area of 519 km^2 of water and sand with 40 km^2 of submerged islands. The *Venezia Nuova* consortium has designed a project of works of titanic dimensions. Renewing the system of dykes and locks is one thing, cleaning them is another! Twenty watercourses collect the water from a basin that is the home of one-and-half million people and flows into the lagoon; this underground water, laden with the active residues of an intensive agriculture, brings about the proliferation of the algae. Polluted sediments have accumulated in the bottoms of streams and canals and something must be done to repair the damage. Can the fragile ecological balance be restored? It is a case of 'saving Venice'.

The measures taken, insufficient and badly co-ordinated as they are in spite of the international conventions, should limit the damage at least. Those interventions which are meant to tackle the immediate objective of preserving the natural coastal environment are of an urgent nature, but their actual application is a delicate matter. The natural parks and reserves multiply. In France, the protected coastal zones occupy 15,000 hectares. Some islands are completely protected, like, for example, Formentera in the Balearics; Gozo near Malta; Skiatos and Skopelos in the Aegean; Mali Losinj in Dalmatia, and the national park of Port-Cros.

The pessimists belief that all these measures will have to face a new menace which is threatening the coastline: the rising of the sea level which can reach as much as 20 cm from now until the year 2025 as a result of the hypothetical warming of the planet because of the greenhouse effect. Together with the phenomena of geological subsidence which has happened in various places along the coast (the Nile delta has subsided by

590 cm in a century), such rising of the water level may prove of great peril. Anyway, one has to estimate the risks of climatic changes in the light of the periodic variations that characterize the Quartenary Age.

THE PEOPLE

How could poetic illusion lead the founder of the French school of geography to write, at the beginning of the twentieth century that 'nowhere else, more than in the Mediterranean, can one trace farther back stable customs and lasting civilizations'? Imagining the peoples of the Mediterranean as living in wisdom and peace around the sea, everybody in his own place, is as contrary to the lessons of history as believing – as indeed did some scientists of the same period – in the existence and the survival of the 'Mediterranean man'. Perhaps it is only among the Egyptians, living in an oasis, can one meet the archetype of permanence and stability.

Invasions, wars, colonizations, expulsions, deportations, and ethnic cleansings have had the better of the concept of a race. From beyond the sea's shores have come the invading nomads. In the basin itself, cross-currents have been experienced for centuries.

The colonizations of the Phoenicians, the Greeks, and the Romans have left their traces; to a lesser extent, so did the expulsion of the Jews from Spain, which made most of them settle on the opposite shores in Muslim lands, and then that of the *moriscos* of Andalusia who 'returned' to the Maghreb. More recently, the colonization projects carried out by the French and the Italians have further upset these already fragile equilibriums, before new expulsions tackled their *pieds noir*, this sort of ethnic group, as Albert Camus described them. These intermixtures, the result of the vagaries of history, have been experienced in a most dramatic way in the Balkans, as the devastation that followed the dismemberment of Yugoslavia bears witness.

'The Sicilians,' writes Dominique Fernandez, 'have been Greeks, Carthaginians, Romans, Byzantines, Arabs, Normans, Spaniards, Neapolitans, and Italians. Sicilians they have never been, or they have ceased being so for ages! And this is exactly their drama: they do not know who they are ... .' In any case, they are Mediterranean people, will reply the archaeologists who have spent so many years publishing numerous studies to try to prove the unity of this fraction of humanity; actually they have only succeeded to convince themselves since the values they have individualized as being characteristic – the sense of honour and friendship, family solidarity – are actually quite widespread. The man described not without commitment and ingenuity by English and American anthropologist in the Fifties is a terrestrial: an inhabitant of the hills and the mountains who escapes from the rocky or marshy coasts which are almost always inhospitable. He is a farmer or a mountain man who is closer to his relatives of the mainland than with inhabitants of the coastal zones. If there really is a Mediterranean man, as Pitt-Rivers and Davis insist, do we not have to look for him in the port areas? This people of sailors, fishermen, and porters, is certainly not a silent one, but it has not succeeded in making itself heard.

What gets heard, today, are those fractions of humanity who are always on the move. The importance of migrations throughout the Mediterranean basin is of such demographic relevance – but also of social, cultural, and political importance – that it ends up dominating the social dynamics in some regions. In the Middle East, within the fragile borders of the states, the massive presence of 'foreigners' reveals this demographic uneasiness. Up to 3 million Egyptians have been scattered in other Arab countries and 300,000 Syrians in the Gulf States. In Libya, one-third of the inhabitants are foreigners who represent almost one-half of the active population; the percentages are higher still in Saudi Arabia, although the exact numbers are carefully kept secret. And has it not been said that modern Kuwait has been 'made' by 400,000 Palestinians?

Most of the migratory movements that gave rise to this human patchwork takes place within the Mediterranean basin. The centrifugal forces have, however, taken the upper hand (even, in this case, it is the notion of togetherness that is being contested). It was the *alyah* (the diaspora) of the Jews of central and eastern Europe towards the Mediterranean, amongst other factors, which led to the creation of the state of Israel. Emigration to places beyond the Mediterranean has increased greatly. The diaspora of the Palestinians did not only drive them to neighbouring countries; their elite has been dispersed all over the entire world. The Lebanese diaspora is of even older origin. It began towards the mid-nineteenth century and has assumed such proportions that, even before the civil war, there were more Lebanese living in 70 countries than there were in Lebanon itself, where about 3 million live. The channels of this diaspora feed persistent migratory currents and financial movements that are often of benefit to the mother country.

Anyway the majority of Mediterranean migrants is made up of those workers who left the south for the north in the Fifties, Sixties, and Seventies. Two million Italians from the south have supplied the manpower for the industries in the developed part of the country, after the process that had moved numerous Italian families to north-eastern Europe. In the same manner, the Anatolians from inner Turkey move towards the coastal zones and Istanbul without, however, interrupting the wave of migration towards Germany. The combination of a consistent exodus from the countryside with a migration towards foreign lands is also a characteristic of Greece which has more than a million of its citizens living abroad. As in Portugal – one out of ten Portuguese lives abroad – the festivities for the temporary return of these emigrants, during the holiday month of August, are described as 'Mediterranean'. If the millions of Moroccans and Algerians who live in Europe – 9 per cent of the active population – are added, one cannot feel surprised at the total number of Mediterranean people who are working away from their native countries; according to estimates that are in fair agreement, this amounts to 15 million. (It is worth noting that now both Italy and Spain have become countries which are now experiencing the phenomenon of immigration.)

The most evident line of division, still, is that which divides the Mediterranean on demographic lines, that is that of a Latin north from a south which is generally identified with the sigla CSEM – Countries of the Southern and Eastern Mediterranean. In the CSEM the south is made up of the Arab countries, while the east is made up of the former Yugoslavia, Albania, Turkey (with 60 million inhabitants, who make up two-thirds of the entire population of the CSEM), Malta, Cyprus, and Israel. The demographic growth has been spectacular: in forty years, from 1950 to 1990, the population of the 'Latin' countries grew from 125 to 162 million, while that of the CSEM increased from 87 to 219 million. Growth is equivalent to 30 and 150 per cent respectively. Worth of particular note is the change of the distribution of the forces: the 'Latin' countries had 59 per cent of the population of the Mediterranean in 1950, but only 43 per cent in 1990. At present the population of the Latin countries grows at the rate of between 0 and 0.5 per cent annually, while those of most of the CSEM varies between 2.5 and 3 per cent. The Latin countries are, therefore, characterized by weak demographic movements: a low mortality rate but also a low birth rate owing to a decline in fertility. Thus falls another myth, according to which Mediterranean people were forced to emigrate because of their fecundity. The decline in fertility goes back to the nineteenth century and not even the dictatorial regimes succeeded in halting it. The proclamations and the measures in favour of higher birth rates issued by the Duce, the Caudillo, and the Greek colonels did not stop the slow march to a situation that was to see insufficient births to guarantee a proper generational change. The birth rate in Italy is beyond doubt the lowest in the world; indeed, even if it still reaches 14 per thousand in the south, it is barely 7 per thousand in the north-west and in the central regions. For twenty years the generations have not been renewing themselves. The situation is not much different in Spain, a country that has also reached European standards rather late; after a temporary explosion in births, the rate of population growth in Spain is declining fast and has recently been moving towards the lowest European figure of 7 per thousand. Even Greece, always as a result of the decrease in fertility, has found itself in the same situation of insufficient generational change since the Eighties. France does not cut a bad figure in demographic terms with regard to the other European states, even if this is certainly not due to its 'Mediterranean population': it has a birth rate of about 13 per cent, but for twenty years the figure has kept below the fateful level. It is evident that the demographic growth in the Latin countries, in spite of a generally positive migratory balance, has become extraordinarily slow.

The extremely fast demographic growth of the CSEM is rightly considered by most experts as being central to the Mediterranean problem. Its interaction with the economy, the environment, and politics is evident.

Turkey, even more than Egypt, is the member of the CSEM with the largest population. Although its fertility rate has decreased from 6.4 to 3.6 per cent in 1990 as a result of the process of modernization, the annual demographic increase remains considerable: 2.3 per cent. A million new citizens are born every year; this corresponds to the number of new jobs that have to be created, in a country where 55 per cent of the population is under 25 years.

The demographic situation in the desert countries, except Egypt, seems to share contradictory realities of a high fertility rate and a tendency for emigration. The importance of foreigners in Libya and in the Arabian peninsula has already been noted, together with the Arab resistance to birth control. The average number of children borne by each woman of child-bearing age is 7.8. At present, in Egypt and in Syria the figure has gone done to 4.5. In the latter country, however, the prohibition of contraceptive measures in action until the end of the Eighties had brought the figure to a record 8 children per woman. The population has tripled in thirty years and the demographic forecast had been regularly exceeded until education, together with the economic crisis, convinced the authorities to intervene. The myth of under-population, which had made the countries believe in the ephemeral United Arab Republic (Egypt-Syria) and the eventuality that the Egyptian *fellahin* would colonize the banks of the Barada, was then abandoned.

In Palestine, the demographic comparison is very significant. In Israel, the Jewish population is growing at the rate of 1.4 per cent annually, while that of the Arab population is growing at 3 per cent. Even more interesting is the comparison between the fertility rates of the Arab populations: it is 2.5 per cent among Christians, 4.2 among Druzes, and 4.7 among Muslims, while in Jordan it reaches 7.1 and in Gaza, which is an 'urbanized' and 'educated' territory, it is 8.5. In this case, it is not only the ethnic identities which are at stake.

Demographic growth in the Maghreb directly concerns, in many respects, all of Europe and not only that which lies on the Mediterranean. It has been repeated all too often that such development is explosive; actually the countries of the Maghreb, even if they are all involved, although not to the same extent, in the process of 'demographic transition', are the inheritors of an uncontrolled dynamic. In the three countries, the decline in the mortality rate is responsible for a rapid development growth. Tunisia, first of all, has launched a family plan in the early Sixties that has brought down the growth rate to 2 per cent. The country's traditional urban structure, its effective administrative organization, and, above everything else, the social acceptance of the efforts to improve the condition of women explain this success very well. Even in Morocco some years later, a plan was launched to limit births but various cultural factors impeded its success and the growth rate has stayed around 3.1 per cent. It is in Algeria, especially, that the drama of population growth has become a paroxysm; European public opinion, however, considers this situation as typical of the entire southern Mediterranean area. Following the proclamation of independence, the growth rate has leapt forward and the population doubled in 20 years. The Third-World and socialist ideology of the regime, favourable to a high birth rate, refused until 1983 a family programme proposed by the United Nations which subsequently managed to cut the fertility rate from 8 to 5. The result is that 55 per cent of the population is under 20 years. In the next ten years the population will again grow automatically by about 50 per cent, and will again double itself (from 25 to 48 million) in 2020. What solutions will be possible then for the problems of education and employment? Of the young people aged between 15 and 24, almost all of whom have attended schools, two-thirds are unemployed. The general rates of employment implicate both the impasse and the social crisis: 42 per cent for men and 5 per cent for women (25 per cent in Tunisia).

Considering such data as a whole, demographic experts have forecast figures which give rise to particular reasons for reflection. From 360 million

Mediterranean people in 1990 there will be 420-440 million in 2000, and 520-570 million in 2025. But in 2000 the Latin countries will only represent 35-37 per cent of the total and just 26-31 per cent in 2025. It is worth noting, on the other hand, that in spite of the economic development that such countries as Turkey, Egypt, and Morocco are experiencing, the gross national product of the Latin countries is still equivalent to about 85 per cent of the total of all Mediterranean countries.

Demographic growth, finally, does not only bring about differences among Mediterranean countries; it also applies in a selective manner in various regions, independently of national borders. In the Mediterranean a process is taking place which is accentuated by urbanization and by a movement towards the coast. Statisticians forecast that, from now until 2025, more than 40 per cent of the population will live in the coastal zones, and that the number of citizens will double itself.

THE CITIES

Their sites, their millennial past and its remains, their function as ports, and the set forms of spatial organization and urban settlement might be the reason why the cities confer to the Mediterranean that unity for which one searches in vain. This apparent unity, however, does not resist contemporary commotions: the north-south division once more separates the relatively regular and ordered cities of the north from the sprawling and undisciplined ones of the south. In the two areas, urbanization still marches with giant steps, both as a phenomenon of the concentration of a growing population and as the irresistible process of spatial expansion. In their different ways, the two huge metropolises of Barcelona and Cairo bear witness to this evolution.

Both quantitatively and qualitatively, urbanization in the western countries of the Mediterranean basin differs from that of the south and the east. Towards

1990, the total population of the seaboard countries reached 250 million, almost triple the 90 million of 1950. It will reach 400 million in 2025. In the western Mediterranean countries, the growth of the urban population has been very rapid between 1950 and 1970, as the combined outcome of the rural exodus and demographic growth: during this period the population increased by 50 per cent. Subsequently the rate of increase has slowed down, especially in France and Italy. On the other hand, urban pressure has not slowed down in the CSEM: between 1950 and 1990 the numbers quadrupled, an annual growth of 3 per cent. At present more than 60 per cent of the citizens of the Mediterranean basin live in cities.

Statistical forecasts made for the Blue Plan estimate that between 1985 and 2025, when the population of the northern countries would have increased by 23 million (from 125 to 148), that of the southern ones would have more than tripled itself: from 75 to 241 million. Whatever the precision of these statistics, the forecast demographic explosion is an urban phenomenon in particular: the threat lies in the cities.

Some more sophisticated statistical studies have confirmed this datum, if the population of the seaboard cities is taken into consideration (and not of the countries): it was about 100 million in 1995 and there were more people in the north (55 million) than in the south (40 million). Everything leads one to think, however, that this proportion will invert itself. According to the Blue Plan, in 2025 the inhabitants of the southern seaboard cities will be 90 million, against the 60 million along the northern seaboard.

The city with the most inhabitants is Cairo with its 9,400,000 citizens, followed by Istanbul with 7,600,000. Of the three western cities, Athens has 3,700,000 inhabitants, while Barcelona and Naples each have 3 million citizens, the same number as that of two southern cities, Algiers with 3,600,000 and Tripoli with 3,100,000. Rome with 2,800,000 citizens and Smyrna with 2,700,000 come close. Another four cities, finally, share the privilege of having more than 1 million inhabitants: Tunis (2 million), Tel Aviv (1,900,000), Beirut (1,800,000), and Marseilles (1,200,000). In all these cities, the rate of growth is quite fast: in Cairo, it is actually explosive. This city, which was already the most populous one in 1950 with 2 million inhabitants, has tripled its population in forty years. Beirut and Smyrna are about to double theirs.

Most of the cities of the Mediterranean are port cities and it is precisely because of this that they have made the history of the Mediterranean world. Antiquity spread in accordance with the rhythm of maritime cities such as Tyre, Sidon, Athens, Alexandria, Carthage, and Rome. In the Middle Ages and in the Early Modern Age, other cities grew during the Arab civilization, including Palermo, Cordoba, and Granada, and, especially, with the development of commercial links between the east and the west, Genoa and Venice both became powerful maritime republics. The displacement of the economic focus towards the Atlantic interrupted this brilliant advance, but only for a period of time. In the nineteenth century, the colonial conquests on one hand and the opening of the Suez Canal on the other offered the ports of the Mediterranean the opportunities of new activities and new wealth. Both along the northern and the southern shores, the port cities have seen the confirmation of their importance, from Barcelona to Marseilles, from Genoa to the Piraeus, from Algiers to Tripoli, from Alexandria to Beirut and Smyrna. They have all become industrial cities, dedicated to the transformation of imported products or the manufacturing of products for export, and often characterized by the oil hegemony which spreads petrochemical plants and refineries.

Little by little, with the growth in their populations, these cities have become complex organisms in which industry does not play the leading role, which, incidentally, has not been taken over by port activities again. The question is often asked as to what these cities and their inhabitants live on. As regards the cities of the southern seaboard, in any case, the answer lies both in the distribution of the active population (the tertiary occupies a preponderant role) and in an informal economy made up of small-scale activities of retail and services. All in all, the city receives its share of the gross domestic product and also profits from the redistribution of foreign contributions. The general improvement in the standard of living, which followed economic expansion, together with the spread of secondary and tertiary education, have produced a social evolution. Between the aristocracy, often the possessors of land, and the 'little people', a middle class stratified according to its income, has been gradually formed whose members ranges from clerical workers to business people. The sub-proletariat remains numerous and already the social crisis is involving young graduates, a mass of educated unemployed which is reduced to a state of dangerous inactivity.

From this point of view, even though Marseilles, Naples, and Athens lie within the European Union, they are cities whose ambivalent image is easily associated with a sort of Mediterranean under-development. Today, these three port cities have become important industrial centres, both as regards their internal activity and as regards their steel and chemical plants. Metropolises of both a regional and a national character, they carry out a leading role in various fields. The university and cultural activities of which they are centres confer upon them a well-merited prestige. The investments in urban infrastructure and in communications, although they are never enough, have certainly pushed them into the modern age.

Their image, anyway, remains a fragile one because of their inability to come to terms with a permanent social crisis, linked especially with the problems of employment and housing. Unemployment is a phenomenon which is bound to become even more serious inasmuch as work opportunities attract or keep in the area a surplus active population; the housing problem, because of the lack of particular specific policies, is the originator of segregations which bring about a permanent state of conflict.

The nature of residential settlements translates social stratification into particular spaces. The city centre, if it is not being rehabilitated, has been gradually deserted by the aristocracy who have abandoned their palaces for luxury villas in the suburbs, while the popular classes crowd in the old structures. The respective quarters differ from each other according to the social and even the ethnic characteristics of the population. Residential quarters in modernized centres or in welcoming suburban areas are the prerogatives of the better-off. Finally, the spaces in between and the empty areas have been taken over, according to particular cases, by precarious habitations and, even more often, by houses that lack grace and all comforts. In this way the housing problem is somewhat attenuated without being permanently solved and a situation is created whereby one is pushed or pushes somebody else to make room for newcomers.

Finally, one has to consider that the most serious problems of general order regarding urbanization in the Mediterranean concern the use of space – in a place where nature makes space available in dribs and drabs – and the consumption of water, since local resources are often insufficient in many places, together with the level of air and sea pollution. In this sense the south has little to envy the north where the same problems, although they appear different in a qualitative way, are no less acute or pressing.

The physical relief of the Mediterranean basin, as we have seen, concedes only a limited area to those human activities which need large spaces. In general the topography does not provide extensive plains or at best divides them into small units separated by markedly rugged areas; the pressing mountainous hinterlands do not offer any practical possibility for expansion. Unless one attempts to redress nature, a Herculean task which, however, modern equipment makes possible. This is what has happened with the draining of marshy areas, the construction of roads, tunnels, and canals, or the levelling of the more stubborn rocky faults.

Although the cultivable areas around Mediterranean cities were always limited, in the past they were surrounded by fields from which most of the necessary supplies could be obtained. However, it was precisely these fields that first bore the brunt of urbanization, as can be seen in Naples, in Valencia, and even in the Nile delta south of Cairo. How much of this space is taken over does not depend only on the increase of population, even though this is responsible for the construction of *villas en mitage* in the suburbs, allotments, and collective fields with their network of roads and streets. In the Blue Plan, it has been estimated that the total 'need' for land per unit is 250 m^2 in the northern areas and 40 m^2 in the south! But it is not only demography that is so avid for spaces; industry has demanded and still demands a considerable area of land, both for its large steel or petrochemical plants – often most frequently built on reclaimed barren lands – such at Marseilles-Fos, Genoa, Taranto, Alexandria, and Smyrna, or for shops, laboratories, or the offices of these technocities. The demand for infrastructure, for green spaces for the purpose of recreation, and even for cemeteries, implies a great increase in the use of space, and this is even more felt because the yearly influx along the coast of hundreds of millions of tourists necessitates the expansion of the infrastructure for the demands of a season that, all in all, is relatively short.

The supply of water is another critical problem. The demand for potable water by the urban populations in the southern and eastern Mediterranean regions is only half-met, and the competition for water creates critical situations everywhere. Since most of the resources are situated in the hinterland, now that boreholes have almost exhausted all the underground water supplies near the shore, the priority which the cities naturally claim means that little water is left over for agriculture. The situation particularly deteriorates during the tourist season, exactly when more water is needed to irrigate the fields; the shortage of the supply may bring about a serious crisis, as has happened, for example, in the Tunisian hinterland.

The increased demand for water forces the public authorities to invest ever-bigger sums of money for the construction of dams and canals. Such long-term projects require many years and considerable credit facilities to be completed. When it comes to diverting rivers, regional and international conflicts invariably rise their heads and demand political solutions. In this matter, there are no specific regulations in the Middle East, but even in the European regions some solutions, like that of supplying Barcelona from the waters of the Rhone, need many years to be completed.

Industry is a great consumer of water: the progress achieved in the technology of water treatment will hopefully make them less of a burden on available resources. And cannot such water be obtained from the inexhaustible supplies of the sea? Present-day desalination plants demonstrate the limitations of such a solution: the cost in terms of energy will keep them for many years too expensive to run on a large scale. The transportation of polar icebergs remains, for the time being, the stuff of science fiction.

The third problem of urbanization in the Mediterranean is that of pollution in the air, in the sea, and on land. The temperate climate, which relatively limits the demand for domestic heating, and the frequent microclimatic changes in the weather and the winds should mean that Mediterranean cities do not suffer too much from atmospheric pollution. Notwithstanding this, atmospheric pollution constitutes a real scourge, with the city centre of Athens, for example, suffering from it to a catastrophic extent. Hundreds of annual deaths are attributed to lung diseases directly related to the *nefos*, a sort of Mediterranean fog, while serious damage is caused to the ancient monuments by airborne chemical particles. Industrial activity which, for reasons of economic costs, is scarcely regulated to protect the environment, amply explain the origin of such pollution. Even more harmful are the exhaust gases of the automobiles the concentration of which in small areas, especially in the cities of the south or the east Mediterranean, more than makes up for the relatively reduced level of motorization. As a result, in cities like Rome, Athens, Algiers, and Cairo, it was rendered necessary to enact measures to limit the circulation of private cars in the city centres.

The problem caused by solid wastes does not present itself as it is used to do in the past. What the Mediterranean cities in the poorer countries lack is what is lacking in all countries: the financial means for its collection, recycling, treatment, and, eventually, its being put to a good use. To which one has to add an unfavourable circumstance: the lack of storage space. One, therefore, should not be surprised on seeing most of the Mediterranean shore regularly marred with refuse dumps often running out of control and which constitute so many assaults on the landscape and the environment.

The pollution of the waters as a result of urbanization is even more serious since it affects both the ocean and the sea. The amount of foul water discharged has increased proportionally more than the rise in the population and the capacity to treat it, despite the modern technologies, is not always used in time. The self-cleaning property of the sea, on which many lay their trust, is very far from lacking consequences. The insufficiency of the drainage system and its frequent bad maintenance explain why the quality of the drinking water can be, accidentally and locally, more than dubious. Cholera and typhoid epidemics have fortunately become rare, but dysentery is still common among tourists who, unlike the natives, are not immune to bacterial contamination. The sea, as we have already seen, is the first victim of such pollution in spite of the protective measures in all seaside resorts. But not even these resorts are safe if pollution from domestic, agricultural, and industrial sources were to spread: the phenomenon of eutrophication in the Adriatic in 1988 was an ecological catastrophe which scared the tourists away.

It is therefore not surprising, given this situation, that the Blue Plan which was adopted in 1975 under the aegis of the United Nations Programme for the Environment has placed regulations about urbanization among its highest priorities to protect the sea and the coast.

Two Destines

Barcelona, including its metropolitan area, is the home of 3 million people. As regards our perception of its image, it is beyond any doubt a Mediterranean city; it is, first of all, a great international metropolis, a city of affluent Europe, a northern city. It is the chief city of a province, the capital of Catalonia, but also the capital of a large linguistic and cultural community, the *paisos catalans*.

The Roman town which archaeologists have discovered is not of great importance for the city: a mediocre port, a small and broken coastal plain, and a mountainous hinterland. Its use as a seat by a political power and then its achievement of a hegemonous commercial role in the Mediterranean which multiplies its trades and crafts, confirm its status as a front-line city which Spain's Atlantic vocation had forced into a sort of centuries-long lethargy.

In the nineteenth century, urbanization transformed Barcelona into a great city. The wealth produced by its industries made possible notable urban restructurization (the well-known *Ensanche*) and the urbanistic and architectonic traditions survive as the various façades of the city centre, the outcome of fertile artistic competitions, bear witness.

After the Second World War, industry lead the process of development: a development without any plans which placed great demands on manpower and brought about an uncontrolled influx of immigrants, Catalans at first, followed by citizens from the south. In the 1960s, the attempt to control this influx and to tackle the housing problem brought to light a critical situation: shacks and tents are the marks of underdevelopment. The poor sanitary conditions at the turn of the century meant that there were more deaths than births. The subsequent demographic growth, at least two-thirds of it, was mostly the result of immigration. During this period Barcelona was still subject to Madrid.

The demand for manpower by industry, which towards 1970 provided work for almost half the active population, maintains a strong migratory current. Workshops and workers are taking over the suburban areas, where the building of great complexes is taking place, in spite of the restoration and the extension of the city centre area reserved for the middle classes. It was then that foreign investment started to flow in: the boards of multi-national companies replaced the former leaders of industry, and Barcelona, taking over the obsolete industrial structures, became a city and prepared for the Olympic Games of 1992. It could start to compete with Madrid, with its own economic activity, banking system, cultural originality, and, from 1977, the regained autonomy of the *Generalitat*, offered it the means to do so. The metropolis has its own technological citadel in the surrounding valleys where the micro-electronic, robotics, software, and biotechnology industries are located.

The port, which has been built and rebuilt and to which is annexed a petrochemical establishment, is still the main source of the city's supplies and plays a secondary role in its urban dynamics.

All in all, Barcelona, 'an urban region that rivals Madrid, surpasses Bilbao, and inspires Valencia, has become a cultural capital, as in effect she had been before, which is experiencing growth and yet in an orderly fashion' reveals itself in the last decade of the century as a city with a technocratic vocation, 'desirous to re-establish in the south an equilibrium with the new economic concentrations of northern Europe' (R. Ferras).

Comparing Barcelona to Cairo means confronting two completely opposed cities: a prosperous metropolis of a rich country and an explosive one of a poor country. Two Mediterranean destinies! Cairo is the capital of a great country, the second in Africa in terms of population and the third in terms of its gross domestic product. The city, therefore, occupies an essential place in the Mediterranean and for the Arab world while being a point of reference for the West. Nobody imagines not considering Cairo a Mediterranean city, although, both economically and culturally, the title rightly belongs to Alexandria, the great port city of the Nile delta, with which it is very effectively linked by all modern means of communication.

As the seat of political authority, principal beneficiary of economic development, and the leading beacon of culture, Cairo concentrates the advantages of a great metropolis and also the related costs. Still Cairo is first of all an Egyptian city and belongs to the Egyptians; even though some of its interests, like the Suez Canal, are obviously Mediterranean, its concerns are rather continental. Cairo is the result of an enormous and persistent rural exodus which is brought about because of the excessive density of the agricultural population, hemmed in limited spaces by the desert.

In 1955 Cairo had 10 million inhabitants. Demographic growth has, however, slowed down, without the threat to 'close' the city to new arrivals having had to be put into operation. Still there is an annual increase of 250,000 people, the result of a massive immigration (4 per cent annually between 1937 and 1966) which has favoured a high birth rate. The immigrants of the 1970s were young and illiterate countryfolk, half of whom did not have any professional qualifications, and with the tendency to form large families. Since the decline of the mortality rate was accompanied by half-hearted birth control measures, no one should be surprised at the population increase that resulted, even though in the 1980s the drift of rural emigrants to foreign countries has limited this growth to 2.3 per cent. If this rhythm is maintained, Cairo, which served as the safety valve for demographic expansion, may become a real powder keg: the occasional and spectacular outbursts marked by the ransacking of stores and hotels of the wealthier quarters provide clear evidence.

Officially, more than half the population of Cairo lives below the poverty line. Always according to official information, unemployment among the population of working age is 17 per cent, but must actually be much higher. Cairo is not an industrial city, in spite of the fact that industry provides a fifth of all jobs. It is the services, especially the public ones, that play the leading economic role. However, the wealth of Cairo which sustains the bourgeoisie and the middle classes is primarily derived from capital: from the imposition of taxes over the entire country; from the management of foreign aid, which represented 18 per cent of the gross domestic product in 1990; from the deals and the profits of the stock exchange; from grand fortunes located there; from the incomes derived from the administration of political power; and so on. Ten million people live, directly or indirectly, on these sources of wealth.

These 10 million people, however, need a place to live in. An existing public policy for the construction of social housing, the creation of totally new residential quarters, and, in particular, a feverish building activity for private houses demonstrate the extent of this urban expansion. The city has actually invaded the Nile delta and extends over 30,000 hectares. The remains of the historic city, from the foundation of Qâirah by the Fatimids in the tenth century until the birth of modern Egypt in 1860, cover an area of 400 hectares. Greater Cairo extends 65 km from east to west and 35 km from north to south. Little by little, thousands of hectares of rich agricultural land have been taken over by urbanization or made barren by the removal of surface soil for manufacturing red bricks. The phenomenon of spontaneous or unregulated urbanization does not result in actual shanty towns but rather in badly-equipped confused building lots. It is estimated that they already occupy 11 per cent of the city area and provide a home for 20 per cent of the entire population. Even the cemeteries are taken over by families that moved from the old city centre because of the appalling living conditions there. In the old quarters, the more or less legal addition of one building on another and further constructions on the roofs have led to an excessive density.

It would be wrong, however, to think that the city is being abandoned. The politicians responsible and the experts have launched a plan of public investment which, even though it may not appear coherent or complied with, has had appreciable results. It is true that it is the bourgeoisie, the businessmen, and, especially, the middle classes who are the principal beneficiaries. Ultramodern buildings have risen on the banks of the Nile; eleven new satellite urban settlement areas have been built in the desert. The liberal tendencies of the government (*infitah*, 'the opening') since 1970 has brought about a mad rush in speculative building. This explains the contrast between the *râqî*, the modern and peaceful residential zones, and the *chabî*, the popular, overcrowded, and ones who make themselves heard.

In spite of the unfavourable economic environment, Cairo had managed to avoid the paralysis and the asphyxia forecast in the early Eighties. The emphasis on the modernization of the communication infrastructure has managed to avoid the worst scenario and the bridges on the Nile, the suburbs, and the first underground line have brought about at least what was indispensable, even though public transport still suffers from overloading. Foreign financial aid has made possible the undertaking of works of a pharaonic scale to complete the sewage system that has finally made it possible to carry and discharge all the drainage water away from the city. Finally major efforts have succeeded in supplying 85 per cent of the houses with electricity and 75 per cent (only) with running water.

'Modern Cairo,' writes André Raymond, 'remains an admirable and even a fascinating city: its life and animation that pulsate without ceasing, in waves of numberless crowds, is a fabulous spectacle; the contrast between the modern zones, often conceived with large open spaces, and the old city, so harmonious seen from afar, leaves an unforgettable impression: the majestic beauty of the Nile continuously renews itself....' Cairo, however, 'risks of becoming a banal city ... There still remains the demographic menace that risks destroying the fragile barriers that politicians and experts strive to put up...'

THE ECONOMY

The position of the Mediterranean in the world economy is characterized both by the dependence of most of the coastal cities and by the obvious division that separates the poor from the wealthy counties. The present reality of the southern and eastern regions of the Mediterranean 'is the result of a fundamental contradiction between the processes of internationalization of the economies and the chain of decisions that are taken elsewhere, on one hand, and a growing marginalization within world economic exchanges, without clear prospects of a possible inversion of tendency. In a world based on interdependence, the Mediterranean suffers from a one-way dependence' (H. Regnault).

The rich Mediterranean states, those which belong to the European Union, have a share of more than 15 per cent of the world market, while the poorer countries of the southern and eastern Mediterranean have less 3 per cent. The respective populations, it is worth remembering, are 160 million in the north and 220 in the south. The gross domestic product per capita of the European Mediterranean is five times that of the countries of the southern and eastern Mediterranean (CSEM).

The statistics relative to the gross national product of each country shows a difference that reveals the heterogeneous economy of the basin. Calculated in dollars in 1991, France and Italy each had about 20,000 and Spain, 12,000. Greece and Portugal, with 6,000, and the former Yugoslavia with 3,000 are somewhat backward. These figures, however, are far higher than those of the CSEM, with the exceptions of Israel (11,000) and Libya (6,000). The two giants of the eastern Mediterranean economy are rather distant from each other: Egypt has 600 and Turkey 1,800. In between there are Tunisia (1,500), Syria (1,100), and Morocco (1,000). With a situation of civil peace, Algeria would reach 2,000 thanks to its hydrocarbons.

The internationalization of the Mediterranean can also be seen in the movement of goods, capital, and people. As far as goods are concerned, foreign trade, of which the north is generally the beneficiary, suffers from an alarming deficit in the case of the CSEM. Algeria and Libya are the only two countries with a positive balance of trade. Egypt, on the other hand, imports three times in value what it exports. The imports of Turkey, Israel, and Tunisia surpass by 50 per cent their respective exports. Even the movement of capital generously favours the north: it is not possible to draw any comparisons at all between the foreign investment in the Latin countries and those in the CSEM. Small as they are, these investments in industrial concerns fulfil a significant role in the weak economies of these countries: in any case, they are warmly welcomed everywhere. The profits from emigration and tourism complete the picture of this great window of the Mediterranean on the outside. The remittance of money by workers abroad – in Europe or in the Arab countries – represent an appreciable amount of foreign currency. In Tunisia they equal one-fifth of the country's exports, in Egypt they are far more; in the former Yugoslavia the profits of emigration equal half those derived from exports. Finally, one cannot ignore at all the impact of tourism on the national balances of payment. Once more, such a parameter favours especially the four countries of the European Union. But even in the CSEM, where there is a large commercial deficit, the money derived from tourism manages to make up for most of it: 60-70 per cent in Tunisia, Morocco, Israel, and Turkey.

Economic relations between Europe and the Mediterranean offer the possibilities for a number of projects, conferences, and resolutions; fundamentally, however, they remain at the rather low level as dictated by the free market. The CSEM countries together are responsible for less than 5 per cent of the European Union exports and just 4 per cent of its total imports. And what matters more is that these figures are showing a downward trend. The process of demediterraneanization of the system of exchanges, brought about in particular by decolonization, has meant that in 30 years French commerce with the CSEM declined from 20 to 5 per cent. The recent stabilization of Latin Europe with respect to the CSEM is due only to Italian commercial growth. The exports of the European Union to the CSEM amount to 45 billion ECU as opposed to 33 billion in imports. But the European Union countries constitute more or less half the openings of the CSEM.

The lack of symmetry is most obvious. Various experts insist that it is not only unjust but, in the long term, dangerous. It is the reason why multilateral and bilateral aid is offered. One must, however, note that this aid is rather modest. The Mediterranean is not the principal recipient of European financial aid, and such aid represents only one-tenth of that received per capita in the CSEM, which corresponds only to 1.3 per cent of the gross domestic product. Curiously enough it is Germany which contributes most consistently to the economic and financial co-operation with the seven countries. The contribution of the Latin countries amounts to less than that of the United Kingdom. The United States and the Arab countries have a more important role in bilateral aid. Those multilateral sources of aid (the World Bank and the African Development Bank) are by far the most consistent, and even the International Monetary Fund is always being asked to provide help.

The participation of the European Union as such is therefore quite weak, but this does not mean that it can be ignored at all. A number of co-operation agreements concluded with certain regions of the Maghreb and the Machrek about twenty years ago have been extended and the sums of money increased. Still the persistent shortage of such aid continues to jar, compared to the great number of Euro-Mediterranean talks. In particular it can be compared to the amount of money being voted to compensate for regional disparities by each European country. The interventions by the various financial organizations of the European Union and the supplementary application of Integrated Mediterranean Projects – the IMPs – make available a large amount of public capital in the European countries against which, by comparison, the funds made available to the CSEM pale into insignificance, until the eventual application of the resolutions passed at the Barcelona Euro-Mediterranean conference of 1995.

Tourism

Mediterranean tourism is no recent phenomenon. The first modern tourist guide to give practical information was published in 1600, less than two years after the death of Philip II. It was the work of a merchant from Antwerp and it dealt with Italy. Some decades later, the fashion of the Grand Tour – a voyage in the Mediterranean – began among young English aristocrats. 'The principal aim of travelling,' wrote Doctor Johnson early in the eighteenth century, 'is to see the shores of the Mediterranean.' The real onslaught, however, started much later. It was only in 1931, after many hesitations, that the hoteliers of the Côte d'Azur agreed together to open their hotels in summer. And it was only after the Second World War that mass tourism started.

It is estimated that more than 100 million foreign tourists visit the shores of the Mediterranean every year. The figure – the largest concentration in the world – has more than doubled since the early Sixties. It is basically a summer tourism that is mostly attracted by the sea, and of which Spain, Italy, and France are the principal beneficiaries.

With more than six million hotel beds and at least as many other beds available in other complexes, the Mediterranean has a fourth of the tourist accommodation in the world, with four-fifths of them in the north-western part of the basin. Indeed it is there that most of the tourists from industrial Europe go, making use of a fast and well-tried system of communication and transport. Germany, which has become the leading world tourist market, supplies the largest number. More than 10 million German tourists make their way to Italy, more than 7 million to Spain, and more than 1 million to Greece. The English and the French prefer the Spanish coasts, while tourists from the Scandinavian countries, the Benelux countries, and Switzerland opt for the most varied destinations.

Spain, which opened to international tourism only after the fall of the Franco regime, today lies in the first place of the Mediterranean tourist economy and has passed countries of old touristic traditions such as

France and Italy. Next come Yugoslavia (before its dismemberment) and Greece, which are also developing rapidly. Still the south and the east remain quite far behind: 50 million hotel occupants compared with the 400 million of the north. Syria and Jordan are trying to enter the market; it is only Albania and Libya – and, for the time being, Algeria – which do not share the manna that the income from tourism provides.

The two general types of touristic regions, the rivieras and the lidos, correspond to the two types of coast – the rocky and sandy beaches. The rivieras are to be found on a complex littoral that joins the beaches within the gulfs to the rocky promontories of the headlands and the calanques. The Côte d'Azur is the oldest and remains the most important Mediterranean riviera. If to it are added the Côte di Maures (Saint Tropez) and the Côte d'Esterel (Saint Raphael), it becomes, after Paris, beyond any doubt the most important touristic centre of France and which receives over 10 million visitors every year. Nice by itself – originally a modest Greek port, then a small Sardinian town at the start of the nineteenth century, and at present an agglomerate with 500,000 citizens – has 11,000 rooms in 320 hotels. From Cannes to Mentone, passing by way of Antibes and Monaco (business tourism), the Côte d'Azur has a chain of resorts which benefits the hinterland. Less prestigious and less equipped, even because of the smaller extent and fragmentation of the coastline, the Ligurian riviera does not have any less intensive activity on either side of Genoa: the western riviera (San Remo) and the eastern one (Rapallo, Portofino).

The development of the Spanish rivieras came later, although the growth has been faster. Since the Fifties, the Costa Brava in Catalonia has been aiming at mass tourism with the construction of all the infrastructure and the buildings that this market demands and today it rivals the Balearics in its hotel facilities. The Costa del Sol in Andalusia has experienced a spectacular boom thanks to the building of the motorway and the airport of Malaga. The wall of cement all along the Marbella waterfront, cheek by jowl with luxurious villas, is a symbol of modern urbanization.

With the return of peace, the Dalmatian riviera has profited significantly from tourism, especially from Germany and central Europe. From Rijeka to Zadar and from Split to Dubrovnik, and also on some of the islands of the archipelago, the tourist resorts have been filled to capacity thanks to their competitive prices. Recently Turkey has entered the tourist market and opened resorts on the Aegean coast, around Smyrna, and on the southern coast.

The rocky shores of the rivieras were the first to benefit from tourism, but now it is actually the flat and sandy beaches of the lidos that attract most tourists. At the northern end of the Adriatic sea, the immense gulf of Venice is surrounded by more than 200 km of coast: a great number of large sandy beaches and a rather shallow sea. Constructions have multiplied and the residences have become second houses or resorts for weekend trippers from the wealthy German cities which are only a few hours away by car. For some tine Rimini has offered an almost symbolic image of mass tourism, now shared with Comacchio. The sandbanks that separate the lagoon from the sea have beaches which draw benefit from the nearness of Venice.

More recently, Languedoc and Roussillon have developed a beach 180 km long between the Camargue and the bathing resort of Roussillon. Its development was planned in the early Sixties by an inter-ministerial planning commission: 7 tourist complexes that include 17 resorts rose from the sand and the sea to offer 650,000 beds for an international clientele (of which one-half in camp sites). Situated between the sea and the marsh near Montpellier, the Grande-Motte has become a place of permanent residence for some.

The beach of the Tunisian Sahel has evolved, together with Djerba, into the privileged destination of a European clientele that is so often regimented in package tours. Its hotel complexes have obtained a very important place in the national economy. One should not finally forget, in this listing, the latest arrival among Mediterranean tourist resorts, the beach of Pamphylia in Turkey, east of Antalya and at the foot of the Anatolian mountains. Even the Atlantic resorts of Costa de la Luz, between Cadiz and Huelva, in Andalusia and the Algarve in Portugal can be considered as Mediterranean by adoption.

Finally, over the last twenty or thirty years, the islands of the Mediterranean have become privileged touristic places. Situated far away from the mainland and having accumulated all the inconveniences of isolation and of a deficient infrastructure – which, according to some, adds to their authenticity – the islands have benefited only lately from the progress in means of transport (but numerous airports have since been built) although they are making up for the delay. About 15 per cent of Mediterranean tourists stay on these islands, where a third of the Club Méditerranée holiday villages in the entire basin are to be found. The Balearic islands, especially Majorca, are equipped for the mass tourist market and are by far the most popular island destination. The Greek islands, Sicily, and Sardinia are more popular than Cyprus, Malta, and Djerba. But if some can be developed further to accept more tourists (to the great annoyance of nature lovers), others are over-saturated. Mykonos is the best example of these latter.

As a general rule, these problems connected with the use of space and the consumption of water are directly related to the growth of tourism. Because of the seasonal nature of such problems, finding a solution adds costs that are out of all proportion. Pollution caused by tourists and added to that produced by the locals and by industry has created situations that are very difficult to bring under control. What is at risk is the well-being of the sea, the conservation of which necessitates considerable technical efforts and financial means. Non-biodegradable refuse is partly eliminated by mechanical sweepers which sift the sand of the beaches but the dangerous pollution with medium- and long-term effects is not always identified or controlled.

A tourist economy, therefore, carries a heavy price. Both the state and the local authorities bear part of these costs and intervene with consistent public investment which does not show an immediate return, although the need for them is clearly visible. On the other hand, private investment, which benefits from public infrastructure, is productive, if one believes in its importance and its origin. Huge real estate projects which include tourist ports, golf courses, commercial centres, campsites, and holiday villages attract the funds of multinational finance groups. Mediterranean tourism continues to attract Arab, English, and Japanese capital, and also that from other countries.

The role of tourism in the Mediterranean economy is hard to calculate. Available statistics which quantify its contribution to the gross domestic product are exclusively regional and national in character and include data which has nothing to do with the Mediterranean. Although the income from international tourism for all Mediterranean countries has jumped from 1 to 40 billion dollars between 1950 and 1995, the contribution of the coastal areas is not known. Does tourism vie with industry and agriculture for the first place in terms of its total product? In spite of appearances, this is not certain.

Industry

A significant phenomenon is often ignored: although until the early Sixties the Mediterranean lay on the periphery of the modern industrial economy, in the space of thirty years it has become one of the world's major industrial areas. This is mainly due to the discovery of oil fields, modernization processes and the low costs of marine transportation, the availability of an inexhaustible supply of manpower in the ports and the cities near the coast, and, possibly, to the massive investments that followed particular political decisions. From 3 per cent in 1950, the total industrial production of the Mediterranean countries – excluding energy products – has risen to a little more than 10 per cent. Between 1960 and 1980, the period of maximum growth, the industrial development of the north Mediterranean countries rose annually to 5.7 per cent and to 6.9 per cent in the case of the countries of the south and the east. After 1980 the growth curves have levelled for the northern countries and average 1 per cent, while the growth has remained more or less unchanged in the south. Still, excluding energy, the manufacturing industry of the north clearly dominates: Italy, France, and Spain together surpass 80 per cent.

Obviously the picture changes drastically if one considers only the Mediterranean regions of these countries. The percentage of value added by the manufacturing industry sited in the coastal zones (as delimited by the Blue Plan) is 50 per cent in the case of Italy, 32 in Spain, and 6 in France, as opposed to 100 per cent in Greece and 90 in Egypt. The lack of symmetry between the north and the south remains impressive.

The older industrial centres, that transformed themselves by restructuring, are obviously situated in the north. These are industries of various types and often quite modernized– agro-alimentary, textile, mechanical, chemical, and so on – linked to the city-port complexes of Barcelona, Marseilles, Genoa, and Naples. Most of the great cities of the south have industrialized more recently and they therefore present identical models: Rome, Athens, Istanbul, Tel Aviv, Jerusalem, Algiers, Damascus, Cairo, and Tunis are capital cities where one-half of the manufacturing industry of their respective countries is concentrated. The textile and fashion industries, in particular, have assumed very important roles in some countries because of the foreign investment they attracted in their search for low-paid workers. The profits generated by such products in Turkey and Tunisia contribute about 70 per cent of the countries' exports.

The uniqueness of Mediterranean industry, in addition to its links with the oil economy, lies in the belt of great heavy industry complexes. Most of these were built in the Seventies, after the crisis that followed the oil price hike had led to the cancellation of some steel projects in Calabria and in West Algeria.

Most of these industrial complexes are situated along the coast because their position is linked to port installations, whether they are newly adapted or, more frequently, built from scratch. Steel plants play an essential role in this system with their transformation of raw materials which are generally imported. This belt is now more or less complete: in a clockwise

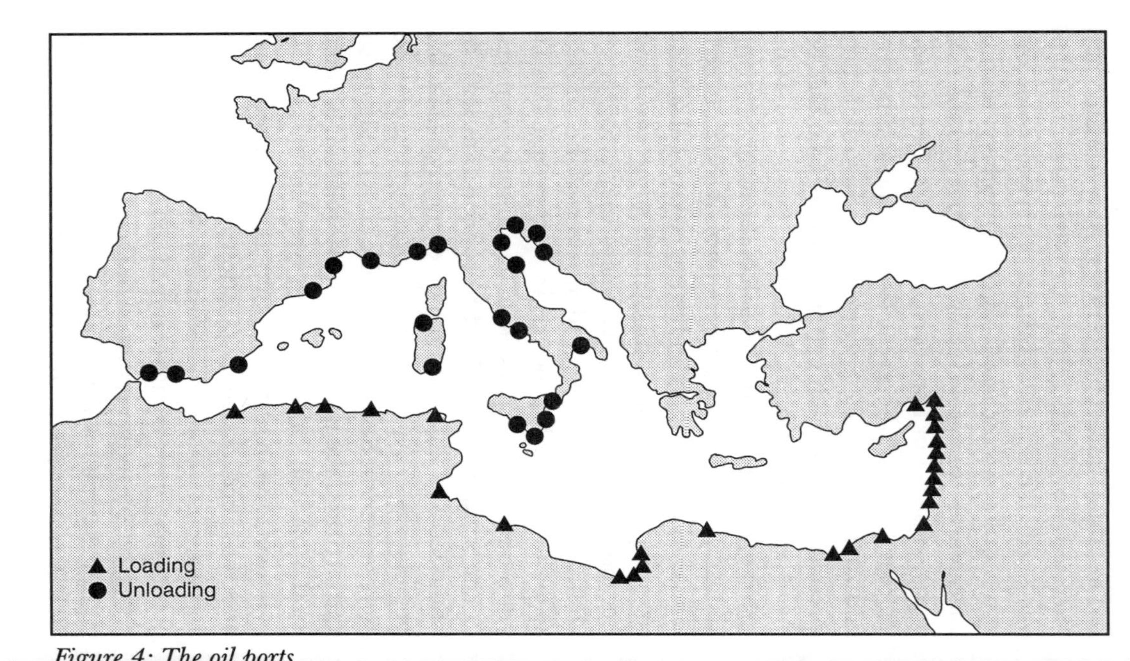

Figure 4: The oil ports

direction there are Marseilles (Fos), Genoa (Cornigliano), Piombino, Taranto, Eleusis, Edremit, Smyrna, Iskendrun, Haifa, Alexandria, Misurata, Menzel-Bourguiba, and Annaba. The Mediterranean steel-making industry, however, has not been spared crises: Italy has closed down the plants at Bagnoli near Naples and Spain that of Sagunto, while the Fos complex produces only one-fifth of what had been originally forecast.

Hydrocarbons, first as a source of energy and then as raw material for the chemical industry, are the key factors of industrial development in the Mediterranean. The production of petroleum and methane in Algeria, Libya, Egypt, and Syria, although important, is less decisive for the economic dynamism of the exchanges between the consuming north and the producing south. Most of the Middle Eastern oil which reaches north-western Europe and North America passes through the Mediterranean. The refineries are mostly constructed on the northern shores, but even the south has equipped itself and the total capacity of the Mediterranean refineries is about 280 million tons of crude oil for 60 refineries, which is equivalent to 8 per cent of all the world's capacity. The evolution of the market, marked by a significant growth in the south, explains why this capacity is not yet saturated; on the contrary, in the north some refineries have had to close down, like, for example, those of Frontignan, Trieste, and Bari. But the added value is not found there any more; it lies in the petrochemical industry, developed, in great complexes linked to the refineries in both the north and the south. Actually the south is tending to develop its own heavy petrochemical industry which is forcing the north, in order to defend itself from the competition, to continue seeking to increase its production in the more sophisticated chemical sector. Consumption takes place outside the area and is concentrated in industrial Europe.

The map of the Mediterranean which shows the refining centres and the petrochemical establishments portrays a scene which goes against what was traditionally imagined. There is one complex after another, all with their networks of supply and distribution by sea and land (oil and gas pipelines). The most important include the Berre basin, Genoa, Sarroch (in Sardinia), Priolo (in Sicily), Izmit, Ras Lanuf, Skikda, Algeciras, Tarragon; to these must be added another dozen of really important ones.

The dangers to the environment from the oil industry are very real, but it looks as if they are under control. There are few drilling operations and offshore stations, and it seems very improbable that concessions for new explorations, which extend over very large areas, can lead to the discovery of deposits like those in the North Sea. The refineries and the petrochemical industries which need thousands of kilometres of pipes and thousands of sluice gates for products which are often inflammable and toxic, create far greater dangers; in actual fact, however, accidents are fortunately quite rare and any pollution they cause is, in the end, less relevant than that produced by normal industrial waste.

Finally, these heavy industry complexes are confronted by those leading industries which have been developed thanks to the attraction of the Mediterranean shores. These are isolated cases, but they are significant since they can prefigure an auspicious future. The Barcelona technopolis represents a remarkable example, and Montpelier, after having hosted IBM, is becoming the place for high-level scientific and technical activity. Sophia-Antipolis, just off the Côte d'Azur is a Californian-type business park. About one hour by car from six inlets, forty tennis courts, the sea, and skiing resorts, about a thousand businesses and laboratories work together in an area of 2,300 hectares, have been built from scratch. The pleasant and interesting geographical position explains why employees, on occasion of some of restructuring of their firms which might mean a transfer to the north, prefer to stay behind and start their own business, benefiting also from the outstanding communications technology.

Agriculture

The importance which industry and tourism have assumed in the economy of the Mediterranean relegates agriculture to the role of a secondary activity, after having dominated until the Fifties. In spite of agrarian reforms, technological advances, and the increase of reclaimed and irrigated areas, agriculture does not manage to satisfy the demand for food of the region and it plays a less significant role in exports. The percentage contributed by agriculture to the gross domestic product of the seaboard countries does not exceed 10 per cent, and even in the CSEM, where it is more than 20, it is in a constant and rapid decline. Nevertheless, the number of people actively involved in agriculture is still rather high in some countries: 40 per cent in Tunisia and 35 in Egypt.

Once more, the north-south contrast is most obvious. By themselves, the Mediterranean regions of France, Italy, Spain, and Greece produce about 60 per cent of the added value in the agro-alimentary sector in the entire basin.

Mediterranean agriculture depends on the area's climatic conditions, but it has adapted itself quite well, although it remains sensitive to extreme variations: the recent years of dry weather, after some exceptional cold spells, have shown considerable differences in their productions. In general, an annual of 400 mm of rain is considered the minimum amount necessary for the dry cultivation of cereals. With a few exceptions, the European regions of the Mediterranean receive more than this amount but on the southern and eastern shores the exception is the rule and only the coastal strips (and the mountains) receive more than 400 mm. Only one-third of the cultivable land in Algeria and Tunisia lies in this situation. It is true that, if the fields have a good capacity of retaining water and if a mild winter makes possible a complete growth cycle before the summer, the growth of cereals easily exceeds the limits and gains ground in the steppes and semi-desert areas.

The climatic conditions, the relations between plain and mountain, the potential for irrigation, the state of the farming communities, demographic pressure, and the eventual intervention of public or private investments all explain the evolution of large agro-pastoral units. The pastoral economy is in decline. The rearing of herds – of sheep and goats – has long linked the mountain and the plain, since transhumance dictated the rhythm of life for breeders and shepherds. But, in general, the difficult search for grazing grounds has divided the farmers from the shepherds rather than drawing them together to seek a complementary economic existence. The fact remains, however, that in traditional farming communities the rearing of ovines is an integral part of the system of cultivation since the fallow grounds supply the necessary land. The great movements of the herds are still rather common even though there is a clear negative trend; however, instead of following the *draille* – the wide tracks over which the herds used to migrate – the animals are today carried by trucks to the mountains in springtime and back towards the plain in autumn.

In the regions which are too dry in summer, the fields of the plains and the tablelands, of the inner basins, and of the gentle hills are used for the growing of cereals. Wheat and barley are well-adapted to the Mediterranean climate. Recent developments have made this particular cultivation, which was for long broken up into small plots and cultivated by means of drought animals, a localized monoculture utilizing the machinery of great investors. The yield, thanks to the use of fertilizers, has improved rapidly.

Elsewhere trees and vines characterize the rural landscape, even if the classic three bases of Mediterranean agriculture – cereals, permanent fields, and the rearing of cattle – have changed greatly. From subsistence economies, arboriculure and viniculture tend to become more and more subjects for speculative investment. The vine, the foremost symbol of Mediterranean civilization, has taken over great flat surfaces just for itself. The replanted olive plantations and, to a smaller extent, those of figs and almonds take over the well-ordered terraces of the foothills.

The most important and significant development, however, has been achieved in irrigation. Small-scale irrigation is certainly a traditional method, and irrigated plots in a particular organized community are certainly no modern invention, from the millennial oases of the Nile valley to the fields around Valencia and the *ghouta* of Damascus. However, the sheer size of the great hydraulic projects that have been completed in the last thirty years, and also the number of smaller ones – like the building of catchment areas and pumping works – is unprecedented. The building of huge dams and the public administration of those lands where an organized 'colonization' is possible have become quite common and already use of much of the available water reserves. France in the Languedoc and Provence, Italy in its *mezzogiorno*, and, especially Spain in Andalusia, have all considerably increased their irrigated areas. More important, in this respect, is the growth of the south. North Africa can yet make notable progress, where development will be significant particularly in the eastern region. Turkey is developing the Pamphylia plains in Antalya and Cilicia in Adana and is in dispute with Syria about grandiose projects to control the waters of the Euphrates. Notice should finally be taken of the pioneering role of Israel in water management projects, thanks to the development of sophisticated methods of drip irrigation.

These irrigated fields supply a market that has diversified itself: rice and maize offer excellent results, as do beet and sugarcane. There is intensive cultivation of crops, including fruit which ranges from apples to avocados, not to mention citrus. Even the cultivation of forage has become profitable, as in the case of Egypt thanks to its herds of cattle. Anyway, the development of irrigation on the Mediterranean has created a number of problems. There is, first of all, a physical problem that is the result of the increased salinity of the over-exploited fields and, in particular, the even more serious one caused by the conflicts that arise out of the conflicting rural and urban demands for water. In spite of the progress achieved in irrigation methods, all Mediterranean countries with the exception of Turkey, are still dependent on outside sources for their food, but this is a sign of the serious lack of equilibrium only in the countries of the south and the east. In most of the CSEM countries, alimentary production per person decreases by more than 1.5 per cent every year, and much more rapidly in years of serious droughts.

The importation of cereals, which can be obtained at cheaper prices on the international market, has doubled in the last 20 years in Egypt and Algeria, tripled in Tunisia, and quadrupled in Syria. In a normal year Tunisia imports half the cereals it consumes, Algeria two-thirds, and Libya three-quarters. Demographic growth and urbanization have placed the problems connected with the supply of food in the first place for those states having to deal with this pressing demand – food shortages have caused popular tumults in both Egypt and Tunisia – by freezing prices of essential food items and by providing hefty subsidies.

In Egypt, one of the world's major importers of wheat, maize, and flours, the price of bread is fixed at a quarter of its value, while the production of wheat, which in 1960 supplied two-thirds of the country's needs, at the end of the 1980s supplied less than 22 per cent. A number of recent measures have doubled production and slowed down the speculative growth in the production of beef and milk destined for the urban middle classes (as a result of which wheat was being grown as forage, with hay coming to cost more than grain).

The countries of the Mediterranean export products which enjoy different fortunes in the international market. In the face of the European demand for fruit and fresh vegetables – a demand which, however, grows rather slowly – Spain has played the major role, thanks to the setting-up a very efficient administrative procedure for exports. The range of climactic conditions, from the Pyrenees to Gibraltar, and the relatively low costs of production, the result of an extremely liberal system which provides a very aggressive commercial set-up for producers who are not that well organized, create a very favourable situation for exporters. Utilizing the administrative procedure that had been originally used for the export of citrus, they have developed a strategy which is particularly effective in the qualitative evolution of demand. Only Israel, which exports a third of its agricultural and two-thirds of its horticultural production, can challenge Spain in terms of efficiency and commercial organization. Turkey exports great quantities of dried fruit. The Maghreb adapts itself with difficulty to the competition while its inefficient commercial organization hinders it from benefiting fully from the advantages of its enviable climactic conditions and low labour costs.

The olive, the symbolic tree *par excellence* of the Mediterranean environment, is in retreat. After long having been the principal supply of vegetal fats in the Mediterranean and the object of long-distance trading since antiquity, olive oil today resists with difficulty the competition of seed oils. In Tunisia, the leading producer of the south, soya oil is consumed more than olive oil! The tree has remained restricted to the poorer areas and there are few regions that specialize in its cultivation: the Tunisian Sahel, the north-western regions of Asia Minor, and Andalusia. Spain is the leading producer and exporter of olive oil. From the European Union, however, it receives generous subsidies and benefits from a whole range of aid programmes, originally set up to compensate small producers, which combine protection and assistance.

Particularly adapted to Mediterranean climate and soil, the vine has been traded since antiquity. The development of viticulture on a large scale and the crisis that followed date back only to the second half of the nineteenth century. Nowadays it is large-scale viticulture which is posing the most difficult problems for the European Union. After Algeria, a 'victim' of viticultural colonialism, had been removed from among the producers, the commercial competition and lobbies limit themselves to France, Italy, and, to a lesser extent, Spain. Fine wines are not at stake. It is table wines which give rise to a competition which is exacerbated by the creation of a single market. Some time after the Languedoc, even Puglia, Sicily, and La Mancha have specialized in this production. In the very unstable table-wine market, Italy has succeeded to increase its exports, benefiting, like the other countries but more efficiently, from subsidies for its distillation which is seen as a solution for general overproduction. It is the contributors rather than the consumers who are paying the price of this wine war. France and Italy have now reached the same level of production, while Spain lags well behind. The other Mediterranean countries do not have much effect on the market, except for some specialized areas, like Porto, the Aegean islands, and Turkey (in spite of the fact that it is an Islamic country), where the pre-1922 Greek population left behind a significant viticultural tradition.

In the end, does the Mediterranean still deserve to be considered 'the garden of Europe'? The very notion of 'early produce', that is of products that benefit from particular climatic conditions, seems to have lost its meaning: the word seems to have disappeared from common parlance. Mediterranean growers are increasingly being forced to market 'even-earlier produce' grown in greenhouses and tunnels that are often heated; they are, obviously, products that do not have the same flavour of those cultivated in open fields and picked at the proper time. And there is an on-going battle with the competition constituted by the importation of ever-earlier produce, not to say out of season produce, like beans from Kenya or melons from the Antilles, or the sophisticated produce of Dutch greenhouses. In a world-wide market, the quality of the horticultural and arboricultural products now comes third, following considerations of price and presentation. Is it, therefore, inevitable that the Mediterranean should renounce its vocation completely? Taste, naturalness, and freshness should be qualities that should still attract customers. But for these qualities to re-assert themselves, there is the organization of production and commerce, not to mention the political will of the European Union authorities, to match the stakes: life or death for Mediterranean agriculture.

THE POLITICAL SCENE

During the second half of the twentieth century, the Mediterranean suffered the greatest rift in its history. This rift has brought about a definite opposition between the south and the north, the latter including Greece and Turkey as well. This transversal rift has shaken the south, fragmenting it like a tectonic movement.

Colonization had for a short time – calculated on a historic scale – assured the Mediterranean of a relative peace during the first half of the century; it was a peace of conquest, of empire, of submission. Decolonization could not be but a process of conflict while the clashes outside the area gave it a dramatic dimension. The inevitable destructurization brings about withdrawal, isolation, and nationalism, and generates instability.

Three main factors explain the original conditions of this geopolitical evolution: the interest of the new Soviet superpower in an area that bordered its own territories; the titanic world-wide east-west confrontation; and the economic and strategic importance of the petroleum deposits accessible through the Mediterranean.

It is in this situation that the south detaches itself abruptly from the north. In the north, restructuring is an on-going process. The setting-up first of a Common Market and then of an Economic Community created the idea of a Mediterranean Europe solidly anchored to the rest of the continent. The successive admission of Greece, Spain, and Portugal has given a concrete dimension to the whole which is consolidated by a commitment to liberalism. Most of the southern states are not able to achieve a political stability to guarantee their internal economic development or to sign foreign agreements to set in motion dynamic synergies. The new independent states, under Soviet and Marxist influence, opt for a state system that breaks away from the area's commercial traditions and turn their structural dissymmetry towards the east. Attempts to unite various states under the flag of the Third World all failed one after the other: the short-lived United Arab Republic, the agreement between Libya and Turkey, and so on. Even OPEC which, thanks to the united front of the Arab countries, had brought about the first oil crisis, broke into a thousand pieces when its monopoly was shaken by the discoveries of new oil deposits. The leadership of Nasser's Egypt and of Boumedienne's Algeria is now a thing of the past. Finally, conflicts between the states (Algeria-Morocco) demonstrate how far affirmations of sovereignty and nationalism destroy the will for international co-operation.

The survival and the expansion of Israel do not even bring about the union among the 'brother countries' to help frustrated Palestine in its search for national identity. The civil war in Lebanon, which ended a centuries-old and symbolic religious and cultural cohabitation, was only stopped by the Syrian occupation. The Gulf War, on its part, has underlined how deep antagonisms can run; inter-Arab solidarity in favour of Iraq under attack was proclaimed only in a few speeches. The Middle East peace process, finally, deepens divisions and cancels the fiction of an existent solidarity. The fall of communism and the international low profile of the former Soviet Union have allowed the power farthest from the region, whose direct interests are less evident, the most ample possibilities of strategic play. In the Mediterranean, the United States act like the world's policeman; to say the truth, also as the real protectors of Israel, their very special ally. In contrast to European reluctance, the American commitment in this unstable region lying near to oil deposits is intended to stop local wars from growing into a general conflagration. It did not stop the division of Lebanon or the dismemberment of Yugoslavia. But American diplomacy, with its powerful means to exert pressure, is determinant in the Middle East peace process and the American army, strongly equipped and supported by the Sixth Fleet, keeps the combatants away from each other in Bosnia. Is the fact that faraway America is today the arbiter of Mediterranean conflicts perhaps a sign, at this time of mondialization, of the extreme dependence of this fragmented region?

Within this overall picture, where every state obviously seeks what it thinks are its own immediate interests, the idea of a united Europe has few probabilities of gaining attention. Still its modesty betrays the fact that its preoccupations, means, and aspirations are all oriented towards the eastern regions of the continent. The Conference for Security and Co-operation in the Mediterranean has set its ambitious aim to open the way for a global systemization of the problems of the Mediterranean basin: it is quite understandable that it is in a state of lethargy.

Words about good intentions regarding indispensable co-operation and unity, answer the resistance of oppressed peoples tempted by Islamic integralism. To the efforts of the intellectuals and artists to create or re-create a collective 'Mediterranean' identity there is a violence which offers them only a choice between imprisonment, death, or exile.

Actually, one could well ask if the energy and imagination used in the creation – never really undertaken – of a regional unity could have been better employed in rationalizing real economic and political assistance for the development of each particular country and region. If the unity of the Mediterranean is a fiction, as we have seen, then it is surely in the interests of the Mediterranean countries to allow themselves not to be entrapped in it.

BIBLIOGRAPHY

Alegre, J., 'La coopération decentralisée', *Confluences,* 7, 1993.

Amin, S., *La Méditerranée dans le monde,* La Decouverte, 1988.

Balta, P. (ed.), *La Méditerranée réinventée. Réalités et espoirs de la coopération,* La Decouverte, 1992.

Birot, P. and J. Dresch, *La Méditerranée et le Moyen Orient,* 2 vols., PUF, 1955.

Blanc, A., M. Drain, and B. Kayser, *L'Europe Méditerranéenne,* PUF, 1967.

Braudel, F., (ed.), *La Méditerranée, l'espace et l'histoire,* Flammarion, 1985.

Burgel, G. *et al., La CEE Méditerranéenne,* SEDES, 1990.

CEFI, *La Méditerranée économique,* Economica, 1992.

CERI, *La Méditerranée, espace de coopération,* Economica, 1994.

Charpentier, B., *La Méditerranée blessée,* Glenat, 1992.

CIHEAM, *Annuaire des économies agricoles des Pays méditerranéens et arabes,* Montpellier 1995.

Commisariato del Piano, *L'Europe, France et la Méditerranée,* La Documentation Française, 1993.
Corm, G., *Le Proche-Orient,* Flammarion, 1993.
Crouzatier, J.M., *Géopolitique de la Méditerranée,* Publisud, 1988.
Davis, J., *People of the Mediterranean,* Routledge and Kegan Paul, 1977.
Deffontaines, P., *El Mediterraneo, la tierra, el mar, los hombres,* Ed. Joventud, Barcelona 1972.
Drevet, J.F., *La Méditerranée, nouvele frontiere pour l'Europe des douze?,* Karthala, 1986.
Dumas, M.-L.(ed.), *Méditerranée occidentale, sécurité et coopération,* FEDN, 1992.
Durand-Dastes, F. and G. Mutin, *Afrique du Nord, Moyen-Orient, Monde indien,* volume in *Géographie Universelle,* Belin-Reclus, 1995.
El Malki, H. (ed.), *La Méditerranée en question,* CNRS,1991.
FAO, *Projet de developpement méditerranéen,* FAC, Rome 1959.
Ferras, R., *Barcelone, croissance d'une métropole,* Anthropos 1977.
French Ministry of the Environment, *L'environnement méditerranéen, contribution française,* La Documentation française, 1995.
Gizard, X. (ed.), *La Méditerranée inquiète,* Datar – de l'Auhe, 1993.
Grenon M. and M. Batisse, *Le Plan bleu,* Economica, 1988.
Isnard, H., *Pays et paysages méditerranéens,* PUF, 1973.
Jeftic, L. (ed.), *Climatic Change and the Mediterranean,*

Edward Arnold, 1992.
Kayser, B., *Les soci étés rurales de la Méditerranée,* Edisud, 1986.
Lacoste, Y. and C., *L'état du Maghreb,* La Découverte, 1991.
F. Lauret, 'Les agricultures méditerranéennes: Nord et Sud, deux destins differents', in *Demeter 96,* 1995.
Lozato-Giotart, J.P., *Méditerranée et tourisme,* Masson, 1990.
Matvejević, P., *Breviario mediterraneo,* Garzanti, 1994.
'La Méditerranée assassinée', *Peuples méditerranéens,* 62-3, 1993.
La Méditerranée dans tous ses états, special number of *Revue méditerranée,* 1990.
Pacem in Maribus. The Mediterranean marine environment and the development of the region, International Ocean Institute – Malta, 1974.
Parain, Ch., *La Méditerranée,* Gallimard, 1936.
Pisani, E., 'Equilibre alimentaire, agricolture et environnement en Méditerranée', *Options méditerranéennes,* series A. 24, 1994.
Pitt-Rivers, I. (ed.), *Mediterranean countrymen,* Mouton and Co., 1963.
Pumain, D., Th. Saint-Julien, and R. Ferras, *France, Europe du Sud,* volume in *Geographie Universelle,* Hachette-Reclus, 1990.
Ravenel, B., *Méditerranée: le Nord contre le Sud,* L'Harmattan, 1990.
Raymond, A,. *Le Caire,* Fayard, 1993.
Regnault, H., 'L'Euro-Méditerranée, impossible nécessité?', *Confluences,* 7, 1993.

Reynaud, Ch. and A. Sid Ahmed, *L'avenir de l'espace méditerranéen,* Publisud, 1991 .

Rhein, E., 'La politique méditerranéenne de la Communauté européenne', *Confluences*, 7, 1993.

Santoro, C.M.(ed.), *Il mosaico mediterraneo,* Il Mulino, 1991.

Siegfried, A., *Vue générale de la Méditerranée,* Gallimard, 1943.

THE SITES

Amigos del Mediterraneo, Madrid.
Association pour la Protection de la Nature, Qayrawan.
Associazione Internazionale per lo Studio delle Civiltà Mediterranee, Istituto Universitario Orientale, Naples.
Centre d'Études Historiques sur la Méditerranée Contemporaine, Université de Provence, Aix-en-Provence.
Centre d'Études Internationales du Maghreb et de la Méditerranée, Tunis.
Centre d'Évaluation et de Recherche sur l'Environnement Méditerranéen, Université d'Aix-Marseilles.
Centre international des Hautes Études Agronomiques Méditerranéennes, Montpellier.
Centro di Studi Mediterranei, Agrigento.
Commission Internationale pour l'Exploration Scientifique de la Méditerranée, Monaco.

Ecomediterrania, Barcelona.
Fondation René Seydoux pour la Méditerranée, Paris.
Institut Méditerranéen de l'Eau, Marseilles.
Institut Méditerranéen d'Études Strategiques, Toulon.
International Centre for Advanced Mediterranean Agronomic Studies, Chania-Bari-Montpellier-Zaragosa.
Istituto di Ricerche sull'Economia Mediterranea, Naples.
Mediterranean Affairs, Washington (publisher of the *Mediterranean Quarterly* journal).
Mediterranean Information Office on Environment, Culture, and Sustainable Development, Athens.
Mediterranean Social Sciences Network, University of Malta.
Turkish Marine Environment Protection Association, Istanbul.
United Nations Programme for the Environment, Action Plan for the Mediterranean: Co-ordinating Unit, Athens; Centre of Regional Activity, Sofia-Antipolis.